AF782880

# TIME CONSCIOUSNESS

# TIME CONSCIOUSNESS

## *The Philosophical Uses of History*

**Gabriel Ricci**

**Transaction Publishers**
*New Brunswick (U.S.A.) and London (U.K.)*

Copyright © 2002 by Transaction Publishers, New Brunswick, New Jersey.

All rights reserved under International and Pan-American Copyright Conventions. No part of this book may be reproduced or transmitted in any form or by any means, electronic or mechanical, including photocopy, recording, or any information storage and retrieval system, without prior permission in writing from the publisher. All inquiries should be addressed to Transaction Publishers, Rutgers—The State University, 35 Berrue Circle, Piscataway, New Jersey 08854-8042.

This book is printed on acid-free paper that meets the American National Standard for Permanence of Paper for Printed Library Materials.

Library of Congress Catalog Number: 2002068460
ISBN: 0-7658-0111-6 (alk. paper)
Printed in the United States of America

Library of Congress Cataloging-in-Publication Data

Ricci, Gabriel R., 1948-
Time consciousness : the philosophical uses of history / Gabriel Ricci
p. cm.
Includes bibliographical references and index.
ISBN 0-7658-0111-6 (alk. paper)
1. Time. 2. History—Philosophy. 3. Consciousness. I. Title.

BD638 465 2002
115—dc21 2002068460

# Contents

# Acknowledgments

This study has a long history and I have received much support along the way. Gerhard Spiegler set the stage when he turned my attention to the historical nature of philosophy and made it possible for me to work with Ernst Troeltsch scholars at the University of Hamburg and Augsburg University. Professor Spiegler also alerted nne to Goethe's role in the emergence of nineteenth-century historicism. I was admonished more than once that unless I read *Faust* in a critical way, I would not understand the origins of historicism, nor could I claim to be educated. After many readings of the Faust story, I am been made aware of the problems associated with making such claims. For those who make it beyond the acknowledgments, you will see how the Faust story is woven into the text.

I am grateful to Hermann Fischer and Horst Renz for providing invaluable archival materials for my research. The University of Hamburg and the Ernst Troeltsch Gesellschaft generously financed early stages of my research. At that time Klaus Kodalle and Henneke Gulzow gave my work critical attention and made sure I did not waste away in the library. Without the assistance of Giorgio Tagliacozzo, I would not have recognized the family resemblance between the philosophy of Giambattitista Vico and the humanist foundations of German historicist thinking. His advice that I turn to the work of Ernesto Grassi made it possible for me to see the connection between rhetoric and the historicist worldview. I must also thank Donald Philip Verene for passing judgment on earlier incarnations of this manuscript.

In the final stages of my work I benefited from discussions with colleagues David Brown, who has shared his historical discoveries and passion for historiography with me, and Paul Gottfried, whose intellectual breadth has saved many trips to the library. I am grateful to Anthony Matteo and Ronald McAllister for arranging institutional support from Elizabethtown College.

Finally, without the perseverance and support of Patricia Likos Ricci, this manuscript would still be scattered around the house. At Transaction Publishers I am grateful for Irving Horowtitz's continued support and encouragement and for the final polishing that Larry Mintz has given the manuscript.

# Preface

When Mephistopheles seizes the opportunity to counsel a student who has come to Faust for academic advice, he is intent on creating another vulnerable soul. He directs the student to begin with logic, and then to proceed with disciplines which will aid in classification, without which the world would not speak to us. The analytic curriculum proposed to the ambitious student is summed up in the advice that the student, if he really wants to understand, must first excise what is living in a thing before its parts can be evaluated. By this time in Goethe's drama, Faust has relinquished his rational hold on life, since it has brought him nothing but despair. His lament that he is tired of digging around in words is a disavowal of the methodical approach to life which, along with the scientific apparatus which surrounds him in his study, was to open the door to truth. Mephistopheles's teasing manipulation of the ardent student is symbolic of the long hold which analysis and methods which reduce entities to ever more discrete parts has had on the intellectual imagination. What Mephisto exploits is the problem for which pre-Socratic philosophy is renowned and which was further inculcated into the philosophical tradition in the achievement of Galileo, and after that by Descartes. It is the proverbial problem of motion, along with methods which spatially categorize reality. Descartes transmitted this tendency to the modern philosophical imagination, which consequently became imprinted with the description of matter as extension and mind as distinctly without extension, since it was described as disembodied. Matter was declared to be circumscribed by space, which both provides a discrete location and separation from other entities. The quality of existence which applies is differentiation.

The influence of pre-Socratic natural science, echoed by Galileo and others, has forged an explanatory framework through which an object's motion is calculated in terms of distance traversed, measured in time. The fundamental presupposition of this attitude is that objects are subject to spatial demarcation and thus take on an immobile character. While this procedure may very well have its merits with phenomena which clearly "occupy" space, it ought not be so liberally applied to the world in which entities are not spatially delineated. This other world is the historical realm which is generated by the productions of human consciousness and in which internal temporal dialogue is the norm. This claim has been made periodically in the Western philosophical tradition, perhaps most forcefully by Giambattista Vico whose innovations promised a rational historical science embedded in what today goes by the general name of hermeneutics.

The scientific model relies upon the verity of immobile objects which can be readily measured. Human reality, on the other hand, is not pieced together in the way one might construct a physical object or dismantle the same to get a sense of its inner workings. This fundamental distinction in method was the essence of Henri Bergson's philosophical innovations. Bergson sensed the methodological consequences of evolutionary theory which linked reality in a continuous chain of being. His conclusion that time could not be readily delineated by separate moments was a corrective of the philosophical tradition which did not take into account the reality of consciousness which endures through time, and which cannot by definition be spatially restricted. As a counterpoint to the analytic posture of conventional philosophy, Bergson's synthetic approach, which justified the incorporation of the inner continuity of consciousness, was not alone in calling into question not only habits of the mind but habits of language which had been pressed into the analytic mould. How consciousness ought to be conceived as a heterogeneous unity, as he called it, or of one piece, was a problem that captured the intellectual imagination of the era. Whether one looks to modernist experimentation in literature and the visual arts, the neo-Kantian revival in the nineteenth century, the new physics, or the rise of depth psychology, one finds the appeal that drew so many to Bergson's lectures–the demand that time no longer be conceived in terms of space and that the mind be temporally construed.

Bergson's admonition has not been received by contemporary philosophers who seek the origin of consciousness within the confines of the brain; delimited by what fits inside a cranium. But the proliferation of theories of mind today derives from a tradition in philosophy which has subverted the historical nature of human existence. There is a tradition in philosophy, however, which has upheld the integral role of historical consciousness in philosophical speculation. Seated in history, this tradition argues, philosophy must come to grips with being embodied in time. This embodiment entails contingency which, on the surface, does not dovetail with the universalist agenda of what has been christened scientific philosophy. The integrity and autonomy of time which forms the basis of Heidegger's and Bergson's philosophy did not get waylaid by contingency or relativism, they equally advanced a metaphysical approach with time as a transcendental medium, along the lines of Kant's proposal of time as a fundamental intuition which orients the self through a transcendental aesthetic.

This brand of metaphysics illuminated a process of development in which succeeding stages of growth are internally fused and which do not appear as one moment juxtaposed to another. How this totality of development was worked out by philosophers who recognized the integral role of history in philosophical speculation is the subject of this work. The internal modification of consciousness through which the emergence of the new was possible had to distinguish itself from the mere realignment of pre-existing parts. This was Bergson's agenda as it was the core of Martin Heidegger's existential analytic. They were, then, occupied with what modern-day cognitive scientists call the "hard" problem, as opposed to the "easy" problem of mind. That is, the problem of consciousness itself, not the elucidation of mental functions which can be readily replicated by computer hardware and easily explained by input mechanisms which easily map onto the objective world.

The proliferation of theories of consciousness today are hamstrung by a materialism which accepts the fundamental presupposition of scientific philosophy, that is, the conflation of time and space. Cognitive scientists who are devoted to the "hard" problem are looking for mind in space, in the direction of the brain which is housed inside a rather confined area. These cognitive scientists hope to explain how consciousness emerges from the activity of synapses, neural excitation, and chemical washes. Faith in naturalizing the mind derives from reductionism and the greater investment in an overarching principle which will explain everything. But if time claims its independence from space, mind cannot be reduced to the juxtapositioning of matter to matter. The indivisible continuity disclosed in the autonomy of the temporal guarantees that succession or development must be measured in terms of the "uninterrupted prolongation of the past into the present which is already blending into the future." (Bergson, *The Creative Mind*, p. 32). This was Bergson's call for a new way of thinking about consciousness. He was not only illuminating the problems associated with understanding consciousness, he was also questioning the conventional manner of describing the inner life. As James Joyce, and others would show, a linguistic approach is required which will transcend the mere juxtapositioning of words to words, if we are to accurately capture the vitality of temporal consciousness.

Joyce had become so closely associated with the linguistic experimentation at the time that Judge Woolsey, who formulated the legal opinion lifting the ban on *Ulysses* in 1933, devotes considerable attention to Joyce's literary style. As Woolsey discerned, Joyce provided a view of not only what his characters saw but an analysis of what was going on in their minds all the while. Joyce's characters, Woolsey noted, were not just occupied with the moment, their attention was focused on a "penumbral zone" equivalent to Freud's unconscious. Consequently, this shadowy region was not so temporally distant if it could appear in the same mental frame as the present.

Joyce was quite clear that Stephen Dedalus was occupied with the problems of historical consciousness. Stephen's desire to wake from the nightmare of history in the opening pages of *Ulysses* reflects both Joyce's difficulties with his historical heritage, and his desire to overcome the ballast of his past. Joyce had previously worked out a solution for Stephen's difficulty with time in *A Portrait of the Artist as a Young Man*. His solution embodies the dynamic in Bergson's conception of duration and replicates the temporal structure of human existence in Heidegger's *Being and Time.* In the April 6th entry in the diary at the end of *A Portrait,* Stephen expresses the temporal foundation of consciousness which was first articulated by Augustine in *The Confessions*: "The past is consumed in the present, and the present lives only because it gives birth to the future." Stephen's realization follows an epiphany in chapter IV. This illumination is triggered by a sublime experience in which all ages become one for him. This temporal totality follows his encounter with the young woman who appears in the guise of an angel. In an extended metaphor rooted in the transfiguration of Jesus Christ, Joyce illustrates that Stephen's call to the artistic life will involve using his past over again. How this is possible in the first place must mean that from a vantage point in time it is possible to meaningfully incorporate all the facets of time. Stephen discovers a transcendence that the dogmatic teachings of the Church could not provide, that the political history of his beleaguered island could not supply and which his mother tongue could not speak. Following his ecstatic encounter with the young girl, he becomes free "To live, to err, to fall, to triumph, to recreate life out of life." Like the phoenix who rises from the ashes to live another cycle of years, Dedalus's soul ascends from the "grave of boyhood" and confronts the prospect of freely re-creating his life. Released from servitude to home, the fatherland, and the church, he becomes aware that something new can result from the past. The chronic brooding that occupied him in the past is exchanged for the thoughtful independence that was latent in his mythic name. His shame had been reproduced each time he faced the derision of his schoolmates, but now the power imbued in his past is realized in the primordial truth of his name. Dedalus's history is the history of liberation; it is only actualized when he accepts the legacy of his name and the burden of the remembered past. He is now prepared to "express himself in unfettered freedom," a freedom that comes only when the past is transformed from a sequence of emotionally encumbered events into a meaningful and enduring totality.

Stephen's resolution to the nightmare of history, which still haunts him in *Ulysses,* is worked out by using the past over again. This provides the material basis for his art, but also signifies the temporal agility of consciousness, which can reconfigure the facets of time so that they become a transcendent and liberating medium. What amounts to the fundamental correction of *a priori* rational metaphysics is a reversal of the Platonic scheme of recollection (backwards), since historical existence determines that we can re-collect forwards. Time is revealed to be the self-constituting medium which involuntarily provides the basis of existence. That temporal consciousness is self-constitutive means that it cannot be relinquished in any legitimate attempt to understand consciousness. Above all, this means that consciousness cannot be relegated to the status of a phenomenon which occupies space. Where many cognitive scientists religiously approach the "hard" problem like any other scientific quandary, some like Colin McGinn are not as sanguine. McGinn's reservations are well founded if philosophers and psychologists continue to overlook the fact that the "hard" problem inevitably entails that the object of study and the subject pursuing the study are continuous. Analysis demands the separation of the subject and object of inquiry, but consciousness is not so easily sliced. The "hard" problem is difficult precisely because it anticipates that what is a tool for working out the problem is what is expected to be illuminated by the application of the tool. Gabriel Marcel identified the logic of this kind of problem as a mystery.

It appears that the cognitive scientists are really working on an age-old problem, the proverbial problem of Being. If consciousness is what is behind everything we think, then it takes on the grounding character of Being which is distinguished from entities with only an ontical status. It seems unlikely that such an entity could be subject to analysis, if we understand differentiation to be a fundamental goal of analysis. This leaves the philosopher in the uncomfortable position of not being able to say anything about consciousness, if it has the pervasive or universal character of proverbial Being. This problem did not mystify the authors of the *Upanishads,* who seemed to embrace the conundrum. With a fervent sense of the ubiquity of Being, the ancients were content with the pronouncement that It (Being or consciousness) made the tongue speak but could not be spoken of, it made hearing possible but yet could not be heard. The problem, this tradition concluded, arises with speech which is responsible for differentiation. Without the tool of language, philosophy in the analytic mode does not make much sense. If, however, the cognitive philosopher begins with the understanding that the "hard" problem of resolving the reality of mind is different from clarifying the many functional qualities of mind that are reducible to language and information, then they too must be committed to more than something beyond analysis, which tends to separate and differentiate. Separation assumes the spatial regulation of reality. Consequently, the working out of the so-called "hard" problem requires solutions which look beyond the spatial categorization of reality. There is a full-bodied tradition in the history of philosophy which has addressed the temporal footing of existence that could augment the analytic approach.

There is little doubt that Descartes belongs to the tradition which has fostered the conflation of time and space. Under his influence philosophy has adopted the distinction between material extension and mental life. The historical orientation of life, which Descartes dismissed out of hand, demonstrates that we are *temporally extended* and therefore theoretically in touch with the totality of our being. Joyce represented the salutary nature of time in the character of

Stephen Dedalus, who was able to rise above the prison of his homeland, his religion, and even his language. Like the mythic Daedalus, he discovered the medium through which he could alter the course of Nature's laws. Stephen's epiphany revealed that he could reconstitute his history, not in the revisionist sense of including material that had been excluded, obscured, or distorted; but in the sense that what has transpired is finally appropriated in an act of interpretation. He discovered that the past was not over and done with, a petrified monument with only a life in memory, but raw material which formed the basis of things to come, the foundation for aesthetic being. Chapter V of *A Portrait*, which is replete with a discussion of Thomistic aesthetic principles, demonstrates the efficacy of these revelations.

Stephen's liberation is a literary representation of the temporal duration which Bergson uncovered as the foundation for life. Stephen's discovery that his past was not over and done with, but could be used over again, is the awareness that consciousness is not realized through the cumulative effects of the past, which can be mathematically represented as a sum of essentially immobile phases occupying separate locations. Stephen's final insights verify that consciousness always includes memory which generates anticipation. Memory that is forever emerging as the new, not lifeless thought which has taken up residence in an attic where old things are stored. Memory never grows old like the discarded toys from our youth, it is perpetuated in a looping movement that continually reintegrates us it into the present.

Heidegger based his criticism of the romantic hermeneutics of the influential Leopold von Ranke on such an insight. According to Heidegger, and others before him, Ranke missed the real point of history because he was so preoccupied with uncovering what really happened. Creating a veritable movement in historiography on the basis of his famous dictum *wie es eigentlich gewesen,* Ranke dismissed what his critics viewed as the dynamic quality of history and the integral role of the consciousness of the historian, which could not be denied. Following Windelband's and Rickert's analysis of the value-relatedness of all historical inquiry, Heidegger would ontologically embellish their concept of *Wertbeziehung* into a comprehensive analysis of time. Heidegger would verify Joyce's sensitivity to the temporal constitution of mind and Bergson's appeal to intuition for an understanding of duration, anointing *Dasein* as the paradigmatic historical reality.

Heidegger's work is a philosophical culmination of the critique of Ranke which began with some of Ranke's politically active students in mid-nineteenth century. What they uncovered, and Heidegger articulated, was the function of critique in history. While his work in *Being and Time*, especially, is obscured by idiosyncratic language which does not easily translate, it is clarified by the adoption of the Rankean critique, his absorption in the revival of Kant at the end of the nineteenth and beginning of the twentieth centuries, the inheritance of the concept of intentional consciousness from Brentano and Husserl, and his allegiance to Goethe's concept of Care as portrayed in Book II of *Faust.*

Heidegger himself did not invite an examination of his many influences. He seemed to be purposefully obscure, perhaps because he was ambitious about appearing original. In his later work he sought an identity with the ancient Greek world that some commentators link to his foray into Nazism. In spite of himself Heidegger's clarification of the role of history in philosophy remains intact. This is so because of the integrity of his influences; his political ambitions notwithstanding. While Heidegger features prominently in this work, it is because he is the culmination of a certain brand of historical philosophy. To the extent one appreciates Heidegger's historical formulations, to that degree one must acknowledge the shoulders upon which he stood. I try to sort out some of these antecedents. Some are obvious as in the case of Ernst Troeltsch whose monumental work *Der Historismus und seine Probleme* (1922) is a sweeping review of the rise of historicist thinking in the nineteenth century. Other influences are companions to Heidegger's insights. Vico's contribution to historical philosophy is included because his was the first voice to express alarm with the hegemony of Cartesianism. Like Heidegger, Vico revealed the poetic disposition of human existence upon which human cognition understands reality on the basis of what we can call a personal corporeal constitution. This tendency meant that the traditionally historical oriented disciplines, rhetoric especially, could not be displaced by unadulterated rational methodology.

Lakoff's and Johnson's recent contribution to the debate in philosophy of mind, *The Philosophy of the Flesh*, corroborates what Vico and Heidegger had already sorted out. Lakoff and Johnson's work is an extension of their research on the primacy of metaphor as an innate interpretative tool. They promise a revolutionary corrective of cognitive science, which has suppressed the embodiment of mind. The proof of such an embodiment is the prevalence of our metaphoric interpretation of reality. They make their case without a single reference to Giambattista Vico, who employed a similar but more elaborately conceived theory to argue against the premier purveyor of disembodied consciousness, René Descartes. More efforts like Lakoff's and Johnson's will undoubtedly come along, though the eliminative materialists, the most severe materialists, will be pulling in the other direction. But, for the promoters of embodied consciousness to be comprehensive, they ought to recall the tradition in thought which has always acknowledged the vital role of history. Because history has been valued as the seat of thinking in this tradition, the body is guaranteed a pivotal role. But it is not simply the body in a confined locus that is confirmed when history is disclosed as the province of thinking. Consciousness is not circumscribed by the events of world history, this would simply result in a linear sequence of connected events. Historical consciousness, and this seems to be the most ambitious theoretical insight of modernism, was the awareness that the past did not stand in isolation from the present, nor was it distant from the imagined future. Both Nietzsche and Goethe held similar positions. Among all the pioneers of historicist thinking, Friedrich Meinecke, in *Die Entstehung des Historismus*, valued Goethe's sense of individuality as emblematic.

This temporal consciousness did not elude the generation of English writers who came before the outburst of modernism. Though Charles Dickens and H. G. Wells did not alter the conventional narrative structure to accommodate consciousness's historical flux and internal dialogue, their subject matter anticipated the time-cult that Wyndham Lewis and H. Stuart Hughes perceived as the intellectual core of modernism. The time traveller in Wells's *Time Machine* learned that the best way to predict the future was to invent it. Similarly, Scrooge's agonizing forays into the past, the present of others, and the future inspired a moral desire to live simultaneously in each. Maintaining a narrative genre that future modernists would deem inadequate to express the fluidity of human historicity, Wells and Dickens presaged an historical aesthetic that would displace the passivity of empathetic romanticism. They had already discovered the internal temporalizing that Martin Heidegger would identify as the Care structure of *Dasein*, that Bergson would label the heterogeneous unity of the flux, and that modernists would try to represent in the visual and literary arts through the elimination of space and connective linguistic mechanisms respectively. What Alfred North Whitehead would characterise as the "Fallacy of Misplaced Concreteness" reiterated the philosophical problem disclosed in the discovery of time as the foundation of consciousness As Wyndham Lewis expressed in his volume *Time and Western Man,* reality is the manner in which things run into each other, not where they stand in a discrete concreteness. When thus the subject runs into the object, then the epistemological construal of the relationship must be adjusted. Whitehead, like Heidegger, introduced the concept of concern, which he borrowed from the Quakers, to describe this realignment.

In a final note on Goethe's *Faust* we should acknowledge that the naive student in Faust's study expresses some reservations about Mephistopheles's advice. In the end, however, he is seduced by the devil's final promise that the prescribed regimen will result in knowing the difference between good and evil. The analytic methods underpinning the student's curriculum, as evinced in the final promise, suggest a sharp demarcation between an hypostasized, objective realm and the student's subjective consciousness. As in the religious account upon which the drama with the student is based, such an awareness is limited to the mind of God. Philosophy is seduced by this same conceit whenever consciousness is stripped of its historical connections. This tendency is pervasive in the cognitive sciences as witnessed in eliminative materialism and E.O. Wilson's recent reworking of his sociobiological thesis, which anticipates that cognition will be translatable into circuitry, and creativity understood as the excitation of particular areas of the cognitive apparatus. This outcome will mean that phenomenological theory will be finally supplanted by a comprehensive neuronal mapping of the brain. Jerome Bruner's culturalist theory, on the other hand, is more typical of the constructivist approach in the cognitive sciences. Bruner's recent work looks for mind in the opposite direction of the materialists, who are focused on the brain, since it incorporates a social aspect, which comes with a curriculum that Bruner refers to as the teaching of the three P's: the Past, the Present and the Possible. This obvious historical pedagogy corresponds to the temporal consciousness addressed in this work.

The historical philosophy explored in this work is an alternative to the radical delineation of the epistemological subject and object. Whereas science tends to objectify the world, the continuity of historical consciousness illuminates the moral and aesthetic subjectivity of existence. The consequence is a historical idealism in which subjectivity is imbued with the responsibility of establishing a teleological coherence in which the past is repeatedly reintegrated into the present. This vision uncovers a historical understanding that can no longer be limited to the past, as the conventional subject matter of history, but which encompasses our practical involvement with the present and our anticipation of the future. Consequently, consciousness is temporally structured as to make each of us instinctive autobiographers.

# Introduction

## The Historical and Poetic Imagination

Traditional metaphysics is inimical to the autonomous reality of the sensuous world. The many variations on the transcendental argument make this evident. Whether we look to Plato's bifurcated world in which the sensible world is a mere reflection of the world of ideas, Kant's metaphysical schema in which the categories of the mind synthetically compose reality, or even modern language philosophy's tendency to "semantic ascent," it becomes clear that conventional metaphysics demands that the sensuous and corporeal aspects of existence must be managed by formal conditions and properties, without which the sensuous is so much unintelligible determinateness. Consequently, sense impressions are reduced to sense data, the sensible world becomes a mere shadow of timeless ideas, and the logical properties of natural language must first be sounded before we have a sense of the features of reality. The tendency to rationally ground experience has seduced philosophy primarily because the rational has perennially been identified as the quintessential human attribute, which not only establishes human intellectual sovereignty but which, more significantly, artificially elevates humans beyond their primordial biological setting. But, this tradition has obscured efforts to acknowledge the pivotal role that the historical imagination has in grounding experience. The historical, which is determined through the sensuous orientation to place, has been relegated, by reason, to the status of an ephemeral dynamic which compulsively seeks rational bedrock to ascertain the reality of this very dynamic.

Rationalism purges the sensuous, only to return more assuredly to the sense world in which reason originally received its impetus. The primary aspiration of this approach has been to carve out formal structures of cognitive activity in order to guarantee truth claims.

In this light, the expression "metaphysics of the sensuous" sounds paradoxical. But rationalism's design to filter the sensuous apprehension of reality conceals a deeper problem. As a case in point, language philosophy's belief that honing natural language and sharpening its essential logical properties will provide a more accurate representation of reality ignores the fact that there must first be some consensus on what the reality that language reflects is all about. This circularity is problematic, and language philosophy overcomes the conundrum by making the restructuring of natural language a goal in itself. There sometimes appears to be no appearance of addressing the ontological reality of facts and a world beyond the linguistic. The architecture of language supersedes the ontology of fact. Kant's Copernican Revolution is somewhat of an exception, since time is posited as the primary intuition that orients the self. Filtered through Martin Heidegger's existential analytic, as we will see, temporality establishes an internal transcendence which challenges the validity of conventional extra-historical metaphysics.

Kant turned the philosopher's attention to the mechanisms of cognition as presuppositions for the possibility of experience. Heidegger would transform the Kantian form of transcendence into an analysis of the existential and temporal apparatus which pre-forms experience. Heidegger's use of Kant, then, resulted in the shift from the human modes of cognition, as the foundation for experience, transcendence and ontological knowledge, to the concern for the human mode of existence as the key to understanding the conditions for experience. Charles Sherover's study, *Heidegger, Kant & Time* (1971), demonstrates that this fundamental mode of human existence is the transcendental temporal ground without which the world would not be amenable to understanding. The neo-Kantians, who were primarily concerned with an historical mode of analysis, are not treated in Sherover's work. While Sherover concentrated on the direct link between Heidegger and Kant's temporal schemata, I am chiefly concerned with the neo-Kantian basis of Heidegger's philosophy.

The most damaging maneuver against the sensuous and the historical, after Plato, must rest with René Descartes. His methodical doubt required that he imagine himself a disembodied ego. While his exercise momentarily denied the corporeal, it nonetheless established the certainty of his existence, his mental existence. Descartes' dichotomous world was a more deceptive stratagem than Plato's bifurcated world, since his skepticism promoted a wholehearted negation of the corporeal world. Cast in terms of Hegel's analysis of alienated self-consciousness, Descartes qualifies as a Stoic, since he has essentially defined his consciousness as that which thinks. If there is an otherness in the stoical world of Descartes it is thought-constituted. Thought supplants actuality, and the only determined reality is the self-identity of thinking. Like Hegel's stages of Stoicism and Skepticism in the *Phenomenology of Mind,* Descartes' ascent into the mind institutes the caprice that eclipses the contingency of existence. Like the stage of Skepticism in Hegel's *Phenomenology,* Descartes established that his thinking existence depends upon no material thing. Equally, he strengthens the proof for his existence by first pretending he has no body and that there is no world that he is in.[1] He uses the material world only to discard its shell after it has been diagnosed as trivial to thought. Hegelian Skepticism contains the same bold contradiction; that is, while this form of self-consciousness is occupied with proclaiming the unessential nature of external reality, it affirms the essential validity of its own pronouncements.[2] In exalting the mental core of his reality, Descartes negated the actuality that he must consequently prove if he wants grist for his thinking self. Descartes' disembodied ego is symbolic of rationalism's denigration of the corporeal compass of existence. He may have simply been catering to the educational tastes of the time (particularly the Jesuits with whom he studied and learned to appreciate the militaristic rigors of Ignatian meditation), but he sowed the seeds for a scornful attitude toward sensuous reality that persists to this day, though it has been challenged from even certain quarters within analytic philosophy. The most systematic effort from within the analytic school is the monumental work *Philosophy in the Flesh* (1999) by George Lakoff and Mark Johnson, who have for some time demonstrated the metaphoric, and hence figurative, predisposition of mind. They ignore, however, the tradition in philosophy that has challenged Cartesianism and the unadulterated construal of mind from the outset. Descartes' dualism would receive its most ardent critic in Heidegger who purposefully avoided both the use of the expression "man" and "consciousness" in order to reveal the simultaneity of human existence and being-in-a-world.

At the same time, Descartes promoted the cognitive basis of consciousness to the exclusion of the corporeal, Giambattista Vico devoted himself to heading off this disequilibrium in philosophical thought. Vico was aware of the vicious circularity in this metaphysical approach, and his immersion in speculation on the origin of language and civil institutions cultivated a deep appreciation for the historical role of the figurative. Descartes may have sentenced human existence to a perverted self-estrangement that tantalizes the mind with the shadow of a determinate reality, but, as quickly, Vico rescued the corporeal imagination from self-imposed exile. Vico's work advocates a metaphysics of the sensuous, and Hegel's most infamous critic, Marx, undoubtedly prescribed the same restoration of the sensuous to philosophical speculation.

Marx's effort to correct the alienation in Hegel's speculative philosophy rests on the capacity of feelings and passions to disclose the ontological core of human existence. For example, in the *Economic and Philosophic Manuscripts of 1844* Marx begins the section "The Power of Bourgeois Money" by differentiating the merely anthropological reading of feelings and passions from construing them as ontological indices of our essential being.[1] The characteristic manner in which feelings and passions affirm objects of sense is the sensuous dynamic entailed in Vico's postulate that "man is the measure" of the universe. That is, the sensuous apprehension of an object, especially when it is transformed through practical activity, is a genuine affirmation of the object. In this process the object is not reduced to an object of consciousness, its determinate reality is confirmed as the ontological net of feeling is cast over the object. Hegel's logic, on the other hand, reduces the object to an object of consciousness. Marx, exhibiting a common attitude of radical social reformers, proceeds to rail against the detrimental power of money. Money's ability to make us all things sabotages any effort to actualize the real virtues and strengths of the individual. Our actualization is denied as long as money mediates our existence; our relations with the world and with our fellows, because they do not originate in expressions or perceptions that correspond to the object of our will, denies the possibility of "real individual life." [2] The natural relations that promote the reciprocity of the expression of passion and the actuality they seek to affirm are concealed when money appropriates my existence. Marx, like Vico, uncovered the natural relations between passions and their objects, which underpin our sensuous assimilation of actuality. Human existence estranges itself when money is permitted to exclusively mediate the relations between need and the object of need, when money is allowed to mediate between life and the means of life, and when the existence of others is mediated by money.[3] In becoming all things through the medium of money, personal actualization is denied, personal capability atrophies, the human essence projects the money essence. When money proves that what were handicaps can become the source of creative power, the productive core of human reality is concealed. With the bold assumption that a human world is possible, Marx challenged the nefarious way in which money alienates us from ourselves. He asked us to consider the formula that our relations with reality ought to originate in a "specific expression" of our real individual life that corresponds to the object of our will.[4] The ontological reality of feelings and passions provide the medium through which reality is sensuously mediated. We possess these strengths naturally and they are involuntarily evoked when we come in contact with a corresponding sensuousness in reality. As long as money dominates the mediation of our existence (or any other system in which an external agent pretends to be the source of our creative power), we remain enslaved by relations that conceal the truly creative power of our passions and feelings. What are customarily regarded as relative facets of diverse social and personal structures, that is, feelings and passions, are elevated to inalienable ontological structures through which reality is sensuously assimilated. The irrevocable distance between the object and the subject in orthodox philosophy is overcome in the intrinsic corporeal medium between understanding and reality; an understanding that is imbued with an applicative incentive which necessarily moves beyond the theoretical disposition of philosophy. It was only natural, then, that Marx would define the human essence in terms of a social nexus that entails the activist dynamic in the natural relation between reality and interpretation. The human essence can only be circumscribed in terms of a collective dialectical exchange; the praxis that defines the human essence overcomes the inarticulate taxonomy of being part of a class, being part of a group, being defined by mute categories. (Theses VI on Feuerbach entails this praxical approach to defining the human essence.)

There have been recent challenges to rational foundationalism; Richard Rorty's project stands in bold relief among those who still carry the torch for analytic philosophy, and Lakoff and Johnson's thesis that mind is essentially embodied promises a metaphoric turn in the cognitive sciences, in many respects the heir apparent to the positivist tradition. But there have been long-standing currents in philosophy that have already met this challenge. This philosophical disposition justifies the expression "metaphysics of the sensuous," since it does not elevate the conceptual over the corporeal and the contingent. This outlook regards consciousness as an inclusive presence which is at ease with its own historicity. The solution the "metaphysics of the sensuous" proposes for overcoming the vicious circularity, produced by rationalism's regression into internal precincts of the mind, is found by matching the corporeal features of the manifold with the corporeal features of human existence. This correspondence alleviates the problem of bridging linguistic entities with non-linguistic entities, mental substance with extended substance, etc. The metaphysics of the sensuous, as it appears in the history of philosophy, illuminates the natural relations between word and object, mind and nature, thought and reality, consciousness and determinateness.

Though modern rational metaphysics has its beginnings in the panmathematicism of Descartes and the analytical methods his writings spawned, the clash between a logocentric worldview and a more practical orientation to conceptualization can be traced to ancient sources. The tension between the two philosophical alternatives is described as a kind of madness by Richard Bernstein in *Beyond Objectivism and Relativism.*[1] Bernstein turns to the principles of hermeneutics to alleviate the tension and potential insanity of the objectivist/relativist antithesis. This tension is as old as thought but is perhaps more disputed today, since the objectivist or modern-day rationalists are aligned with the powers of scientific investigation. This position invokes a technocratic idiom to bolster its claims to methodological perfection. Such a footing, however, neglects the historicist world view, which requires all methodology to begin with temporally bound conditions, and which, consequently, questions any strict correspondence between timeless thought and the determinateness of the external world. The historicist worldview garners its strength from (1) the proposition that developmental contexts are essential to interpretation, (2) the productivity of the subject, and (3) the emergence of the new.

The philosophers treated in this work have availed themselves of the hermeneutical insight that human existence is primordially inclined toward understanding and interpreting the world. Most importantly, the interpretative initiative that originates with existence is complemented by the drive to apply what interpretation grasps. The praxical initiative that is explicit in Marx's philosophy embraces this insight, Vico's postulate that primordial poetic existence apprehended reality through a robust corporeal imagination addressed this concern, and Heidegger's existential analytic retrieved the hermeneutical directive that human existence is categorially comported toward understanding its own completion. All three positions recommend the creative and transformative power of the applicative hermeneutical condition as an alternative to either the absoluteness of the objectivist worldview or the nagging skepticism of relativism. In their own way, each thinker is at home with an inclusive notion of human existence, which is not eager to purge either the contingency of existence, or the historicity of consciousness, or the primordial value of time and place, in a word, the sensuous. For example, Heidegger's idiosyncratic use of the concept of *Dasein*, which refers to human existence, is his way of avoiding the problems generated by philosophy when it begins with pure consciousness as the primary ontological facet of human existence. When, like Hegel or even Descartes, the philosopher begins with unadulterated mind, it is very difficult to advance beyond it into an actual world. Heidegger's *Dasein* is inclusive of the world, history, totality, and above all being-toward-possibility, which is its ontological orientation. The Care structure, which is identical to *Dasein,* especially, entails the self-transcending movement of historicity. This means that *Dasein* need not initiate a transcendental movement, since it is primordially transported by the ecstatic movement explicit in its Care structure. The facticity of *Dasein* embodies the historical unity (totality) of consciousness and object that must first be systematically realized in Hegel's philosophy, and which requires reconstitution in Descartes' dualistic world.

Aristotle presented the first real challenge to rationalist metaphysics. His critique of Plato's conception of the transcendental character of the Good in the *Nicomachean Ethics* offers a paradigm for the pattern in Western philosophy that links Vico, Marx, and Heidegger. Aristotle's evaluation of Plato's Good in the *Nichomachean Ethics* turns on the unity of theory and practice which emphatically establishes that the realization of good is preceded by activity.[2] His challenge calls into question the viability of an independent and self-subsistent Good which, Plato argued, could only be realized (understood) through the dialectic. Aristotle did not dismiss a final goal in all that we do, but it too could only be achieved through action. Aristotle's challenge has been taken up by the historically minded philosophers that are treated in this work, and the phrase "metaphysics of the sensuous" refers to the philosophical effort to de-transcendentalize metaphysics and restore it to an historical context in which practical activity, creativity, and sensuous projection onto reality become the focal points for meaning and understanding. Furthermore, the teleological stamp of Aristotle's metaphysics, combined with the contiguity of matter and form, represent a prototypical developmentalism which, it can be said, was taken up by modern-day historicism.

This paradoxical expression is resolved in Aristotle's work, and later by Giambattista Vico's fervent claims in the eighteenth century in which true knowledge and genuine conceptualization entails the dynamic participation of the historical subject, without whose active role true learning can not take place. Scientism, on the other hand, as in the case of Cartesianism, declares that the subject as inquirer is autonomous from the field of study. The reflexivity inherent in the methodological dictum that the method of inquiry should follow from the nature of the entity that is studied is totally obscured. The explicit enmeshment of historical existence in the effort to clarify that very existence is not obscured in the "metaphysics of the sensuous," it becomes the originating principle upon which metaphysical speculation turns.

E. O. Wilson, in his controversial book *Sociobiology: The New Synthesis*, launched a brand of reductionism which anticipated that someday biology would be able to transpose cognition into circuitry.[3] Representative of some contemporary cognitive scientists, this materialistic reductionism, in Wilson's words, will define creativeness "as the alteration of specific portions of the cognitive machinery regulated by input from the emotive centers."[4] In Wilson's world, psychology and philosophy will have been displaced, but more frightening, Stephen Jay Gould and others have argued, would be the regulatory measures that some might find desirable in producing and/or eliminating certain behaviors. This form of reductionism upholds the Cartesian legacy, since creativity and learning have their seat in unadulterated cognition. They are not conceived as vital connections that contingent corporeal existence spontaneously produces on the basis of what Marx identified as the ontological facticity of passions and feelings. The Marxist alternative precludes reductionism, since at all times the corporeal link between the personal expression of individual life and the other (man and nature) is maintained.

Customarily such challenges to conventional metaphysics are dismissed as nonphilosophical, especially as they elicit the historical point of view and its chronic relativism. To tether the metaphysical to the sensuous, as the expression implies, is to reduce the conventional pursuits of knowledge, the truth, and first principles to dynamic contextual frameworks in which the familiarity and acquaintance with the concrete world forms the basis for speculation on the structure of reality. The introspective tendency on the part of traditional metaphysical frameworks, in which the sensuous is eclipsed in an effort to sound the depths of the rational and cognitive core of human existence, is supplemented by the practical connections human existence perpetually has with reality. Conventional metaphysics assumes the universality of reflective consciousness and argues that the senses, at best, can only provide raw data for metaphysical synthesis. This tendency dismisses the historical origins of consciousness and the ubiquity of historically seated consciousness. Furthermore, as Vico argued, instruction in critical and logical methods ignores the vital pedagogical phase in which real learning takes place. Vico's critique, in his *Autobiography,* was that systematic study was nullified in the zeal to sit in judgment of all matters.[5] The primary emphasis that Vico, Marx, and Heidegger place upon the ontological significance of the poetic, or creative transformation of reality, distinguishes their critique of orthodox metaphysics as unabashedly sensuous. There is some wrangling on lumping these three thinkers together, since their interpretations of the creative ranges from communicative practicality in Vico (in the form of philological certainty arising from human choice), to the productive manipulation of reality in Marx (which restores human existence to its social basis), to the mystical-poetic posture that is at the end of philosophy in Heidegger's later work. One fact, however, remains unassailable. Their mutual critique of conceptually fastened metaphysics claims the priority of the creative in the conscious transformation of reality into the familiar historical sphere of human existence. That the difference in the medium of creative transformation may turn on the distinction between the poetic or real productive manipulation or the poetic as the philological and communicative apprehension of reality is an academic issue. All three thinkers have challenged the conceptual foundation of rational metaphysics.[6]

Those familiar with Heidegger's later assessment of the poetic dwelling of human existence, and Vico's postulate that primordial human existence schematized reality according to poetic universals, recognize that the poetic, in the context of human historicity, refers to the creative impulse that always transposes reality into the historical *top*os that is human existence. Heidegger's technical use of the expression *Dasein* entails the criterion that human existence is historically seated in a temporal unity, and thus history becomes the ontological *topos* or mediating term or place within which consciousness resides. The term sensuous, however, derives from Marx whose critique of Hegel's panlogism also claimed the priority and pre-eminence of the practical assimilation of reality. Applying the term "sensuous" to the brand of metaphysics which calls into question conventional metaphysics is to offer an alternative to rationalist foundationalism and the explication of reality on the basis of *a priori* grounds.

Though the term "sensuous" specifically derives from Marx's efforts to resolve the religious essence into the human essence, restoring it to its practical basis, it is implicitly Vichian, since the wisdom of prototypical human existence is equally mediated through the senses and imagination's corporeal extension onto reality. Vico, in fact, offers a description of the originating metaphysics of this primordial poetic wisdom in paragraph 375 of the *New Science*.[7] The metaphysical outlook that governed primordial poetic wisdom, because it was essentially pre-reflective, could not have been rational or abstract as is the metaphysics of refined intellects. Since, as Vico argued, the existence of the first humans was mediated by strong feelings and imagination, so too would their fundamental world view be permeated with this aspect of their passionate constitution. Vico acknowledges in the *New Science* that it is difficult for us to appreciate this sensuous disposition from the vantage point of self-reflection and the logocentrism of the modern world. He offers some insight, in this same section of the *New Science,* on how we might begin to understand this metaphysical outlook. He asks that we keep in mind the model of the imaginative child, who, like the proto-human, projects their being onto inanimate objects in the act of play. The use of the expression "sensuous" is further justified in Vico's explication of the origin of language within such a sensuously mediated metaphysics. Like Rousseau after him, Vico acknowledged the pre-reflective status of the prototypical human existence, since the first stirring of linguistic apprehension, devoid of the ratiocinative, meant that humans made their own corporeal and sensuous existence the measure of the world. Vico was not advocating a Protagorean relativism, but a topical relationism through which reality is mediated by a vigorous corporeal imagination. Accordingly, Vico claims the primacy of figurative language, since it entails the extension of familiar aspects of the body onto reality. The consequence of this originating metaphysics is that knowledge is mediated by making reality out of oneself, much in the same way that the child transforms the inanimate toy according to its own sense of the living self. Vico's methodological prescription, that any doctrine must begin with the matter of which it treats, requires that metaphysics must begin with the first modifications of the human mind. Although Vico acknowledged the difficulty in understanding the original grounding of this historical metaphysics, since consciousness passed through this initial stage in the infancy of humanity, we are continually afforded opportunities to witness how children apprehend what they understand through imitation. The modification of the original poetic wisdom is recapitulated in the life of a child. To argue that rational thought has superseded this metaphorical apprehension of reality is to assert that poetic speech is not capable of linguistic representation in which correct statements can be uttered on the basis of the coherence of a subject and a predicate. More importantly, Vico's postulate of the poetic utterances of the primordial life of humanity corrected the emotionally distant ego of Cartesianism in which the self is poised against the world in a mirror-like posture that receives the world in (all things being equal, and no evil genie in our midst) an objective representation of reality. Meaning, however, need not be literal to be coherent, and expression need not be cognitively inspired to be understood.

Donald Verene has argued, in his in-depth analysis of Vico's conception of the imagination *(ingenium)*, that Vico's metaphysics most closely resembles Kant's regulatory construal of reason. That is, in respect to the given with which all transcendental philosophy must begin, Vico and Kant emphasize the synthetic function of the mind, which implies a constructivist role for consciousness.[1] In the primordial orientation to reality that accompanies the original efforts to fix the manifold of reality, however, Vico postulated that the synthetic activity of consciousness, in relation to an object, occurs simultaneously with an image. Thus, the problems associated with Kant's understanding that the categories somehow bring the object into focus, after the object has been initially framed by the transcendental aesthetic, are avoided. Verene acknowledges the cunning of Vico's metaphysics, since it recognized the autonomy of the image and did not regard it as mere content for conceptualization.[2] What Verene signifies, in placing Vico among the great metaphysical thinkers, is that Vico's historically postulated metaphysics avers the active participation of primordial consciousness in the construction of reality. The corporeal speech that Vico posits of primordial consciousness, and which fixes the transitory, is implicitly an action since the projection of the body onto reality is what originally made the unfamiliar familiar. The reservations that some Vichian scholars have regarding the affinities between Marx and Vico, relative to their respective ideas of practical activity or production, are resolved in Verene's assessment of Vico's concept of imagination. With both Marx and Vico, the given, in philosophical parlance, is really what is made by (historical) consciousness.[3]

Verene's analysis of Vico's most significant example of this function of the word or speech in the image of Jove also alerts us to the mental disposition that accompanied this originating imagistic impulse. This concern is crucial to understanding the sensuous mediation that precedes thought. The personification of the forces of nature, which Vico recognized as the burgeoning of thought, was actually a response to the uncanny and the terrifying. The act of personification drew the horrible and fearful powers of nature into the sphere of the body, that which is by definition familiar. The transformation of reality is generated by awe which is involuntarily read by the body and thus washed with sensation. This reflexive movement guarantees that the sensuous acquaintance accompanying the primordial experience of which Vico spoke is not simply substance for thought but the actual shape of thought. Notwithstanding the two different historical contexts that govern Vico's worldview and Marx's critique, Vico's epistemology unquestionably resembles Marx's understanding that we are responsible for the shape that the civil world takes. That Vico emphasized the sensuous mediation of reality, in the way that Marx did, is made clear if we consider that the originating poetic projection of which he spoke was activated by fear and alienation.

Verene meticulously refers to the many passages in Vico's text that speak of the pervasive fearful disposition which accompanied the primordial poetic human consciousness.[4] These passages establish the sublime nature of this primordial response to the horrid and terrifying qualities of alienating nature. The significance of the experience of the horrible and the terrifying cannot be underestimated in Vico's work, since the *luci* or burnt clearings among the woods were first prepared by those families that had lived withdrawn among the "sacred terrors" of the forests (NS, 16). Given this disposition and awareness of the original peoples, their first articulation of reality could only have been mediated by the sensuous familiarity of the corporeal. The primordial fixing of reality in this original poetic gesture did not use speech in a mere referential way, since the awareness of the terrifying powers of nature did not simply identify those powers; the primordial poetic impulse "made" those forces sensuously intelligible. The assimilation of reality was not a process of labeling but a process of identification with the alienating forces of nature. The primordial effort to classify the world, then, is conditioned by a profound sense of fear. Since Vico had earlier argued in favor of the strain of ancient philosophy rooted in rhetoric, Verene, in his scrupulous effort to illuminate the awe associated with primordial poetic speech, might have, along with Vico, acknowledged that in the infancy of the human world men were sublime poets (NS, 187). This philological axiom, as Vico put it, was borne of a corporeal imagination that passionately incorporates sublime space to the extent that the original poets would be dramatically distressed (NS, 376). The anointed status of this originating consciousness meant that we should see it as poetic, Vico stressed, since the poet in Greek refers to the creator (NS, 376). The task of poetry, in this light, in addition to the cultivation of virtue and the humbling of humans in the face of the terrifying awe of nature, was to continue to perturb with the invention of sublime fables (NS, 376). Vico emphasizes the productive character of this poetic posture and was so certain that language had its beginning in gestures and figurative manipulations that he interpreted *logos,* the word that above all we associate with the rationalism of the ancient Greeks, as deed (NS, 401). The poetic character of *logos* antedates its rational construal, since speech, from the historical point of view, originated in the sublime gesture attributing senses and passions to physical bodies (NS, 402). The poetic charter for the *logos* receives its most dramatic exegesis in Faust's conclusion that "In the beginning was the deed." Faust's idiosyncratic interpretation is a rejection of reason and an indication that he is prepared to finally take real action beyond the confines of his study.

Vico, then, clearly identifies the role of these originating poets as creative. He acknowledged earlier sources that attribute this tendency to the introduction of religious idolatry (NS, 382, 188); but more to the point, this invention of the gods in response to the terror of the powers of nature was really fear brought on by the manner in which human existence is predisposed. It was not fear of the gods, or fear of others that triggered the institution of religion, but fear that came from within the constitution of existence (NS, 382). The metaphysical voice with which the first poets inaugurated institutions was generated by the dynamic of the sublime, which is equally dependent on the hidden, involuntary capacity to assimilate reality as it is on the powers of nature. In this light, it is the metaphysics of the sublime which is the productive foundation upon which the history of future rational speculation begins. Heidegger equally acknowledged the institutionalizing measure of poetry in his reflection on Hölderlin's phrase "...poetically man dwells," since dwelling only first becomes possible through poetry. As such, poetry is really a kind of building.[5] Heidegger also relied upon the Greek sense of *poesis* (making or forming), which obviously stresses the productive capacity of poetry. Most importantly, the active role of the subject in the process of overcoming the explicit alienation of nature (reality) is not an unadulterated mental process in which the object is transformed into an object of mind (Hegel), but a synthetic assimilation of reality to the corporeal. The powerful emotional make-up of primordial poetic universalizing guarantees that the sensuous appropriation of reality is the basis of the historical inauguration of thought.

Vico's consideration of the value of this original metaphysical outlook is emphasized in his allegiance to the more ancient philosophical discipline of rhetoric. Before the writing of the *New Science* he had already clarified the powers of rhetoric, over and against modern rationalism, to illuminate the real character of the human world. (*On the Study Methods of Our Time*, 1709) This consideration bears on the origin of language, and hence civil institutions, since rhetoric relied upon the authority of topical imagery which formed the context for the primordial experience of fear and awe. If the first communicative gestures were so rooted in specific places, then we might expect that the study of language must also address the influence of *topoi* on the structure of language. Vico, who identifies Dante as the Tuscan Homer in several passages of the *New Science* (NS, 786, 817) and relies upon him for numerous etymological demonstrations, might have cited the very first Canto of *The Divine Comedy* to illustrate the power of both fear and place in the creation of poetic insight. Indeed, Dante credits the fear with which he is overcome in the wilderness he has stumbled into in the first Canto as the impetus for his work. The memory (which was regarded as the primary tool in ancient rhetoric) alone of the dark wood was sufficient to instill fear in him. The model discipline for Vico's *New Science* was rhetoric since its methods relied upon the perspicacity of *topoi* or the mediating terms in arguments that could be summoned to meet the demands of a specific oratorical context. Thought, that is, takes shape from the form inherent to a particular location. The topography of sin in Dante's *Inferno* is a perfect example of the rhetorical practice of conjuring up thought and emotion through the vivid representation of place.[6] In turn, Dante's style proves the Horatian insight *ut pictura poesis*, for his verse graphically depicts the terrors of hell. Dante's poetry literally speaks images which become ensconced in memory.

As the model discipline for the *New Science*, we must expect that Vico viewed rhetoric and *ars topica* as the foundation on which thought first took shape in the face of primordial topical settings, and how it was that an original first proposition was posited. This topical grounding is communicated by sensuously mediated feelings that are so powerful that they transcend the particular situation in which they were generated, and, consequently, achieve a generic status like that of a universal or a class concept (NS, 495). This status is most obviously achieved in the way that the memory is etched with the force (fear and awe) of the original experience, as Dante so effectively demonstrated. These poetic genera, however, are not logically deduced but actually invented on the basis of the terrifying impact that certain *loci* have upon the imagination. Vico intended that the art of topics could shed light on the way in which the mind creates as it takes on new shapes. Vico's stern rejection of the critical philosophy associated with Descartes revolved around the topical formation of the mind. The primary faculties around which Vico's sensuous topics revolves (memory, imagination and invention which are tools of the art of rhetoric) are extensions of the body into a place that encounters the terrifying and alienating forces of nature. The poetic genera that are produced from the inventive practice of assimilating and identifying with these terrifying *loci* is the result of the extension of the body into place. It was on this ground that Vico argued for the primacy of discovery and invention as the foundation of human existence, rather than critical philosophy which can only come after the original creative application of the mind.[7] To overlook the role of invention and discovery in systematic learning is to ignore the historical dimension of human reality, which antedates and is antecedent to the development of critical and logical methods. Moreover, Vico's delineation of the faculties implicit in rhetorical consciousness corresponds to the temporal ecstatic dialectic in Heidegger's analytic of *Dasein.* The manner in which human existence is transcendentally turned in upon itself is illuminated in the historical dynamic that underpins consciousness. It is a re-collective temporal condition that gathers up the facets of time and, as such, asymptotically converges on eternity.

For all the criticisms Vico leveled against Cartesianism, the restoration of the historical integration of human existence stands out above the rest. Vico obviously did not dismiss the *moré geometrico* associated with the modern philosophical mentality, the *New Science* follows the format of such an approach. What alarmed Vico was the exaltation of the mathematical method to the exclusion of the process of human historical development. Eclipsing the historical context of human experience meant turning away from the traditional academic pursuits that could provide the student with the foundation upon which to base critical and logical judgment in the first place. The need for the historical dimension in learning, for Vico, meant that a blanket refutation of ancient methods in favor of the rigorous training in logic would ultimately leave the student without subject matter and with an empty voice.

Vico was certain that the art of topics should form the basis of philosophy, rather than critical doctrines. Before a critical judgment can be rendered of an argument or a thing, he reasoned, we must assume that the mind has already established an acquaintance with the subject at hand. Critical philosophy can hone our logical skills and make judgments more exact, but the art of topics, "because it has the function of making minds inventive" (NS, 498) can illuminate the grounding of thought in discovery and invention. The synthetic powers of mind, derived from invention and discovery, can not be superseded by the rigors of discursive thinking, which in its dialectical play overlooks the creative ground that underlies it. Consequently, the disclosure of the ground of thinking will not be expedited by critical methods.

The affinities among Vico, Marx, and Heidegger go beyond their mutual critique of orthodox ontology, and the body of critical literature that has proliferated on Vico in recent years testifies to this fact. It is Ernesto Grassi's work, especially, which has demonstrated the philosophical compatibility of Vico, Marx, and Heidegger. Grassi was in the unique position of knowing the extent to which Vico's work is prefigured in Renaissance philosophy, and he was the first scholar to take Heidegger, a former teacher, to task on his interpretation of humanism. This eminently prepared him to see the vital links between Vico and Heidegger. Marx dovetails with Vico and Heidegger, as Grassi points out in an essay on these three related thinkers, since Marx equally disputed, *vis-à-vis* Hegel, the speculative vision that derives (historical) reality *a priori*.[8] The connection Grassi draws between Vico and Heidegger, in this essay and in his other works, stresses the weight both thinkers place on the emergence of historical reality by virtue of the imaginative activity of poetic consciousness. Grassi argues that long before Heidegger situated the space of the institution of history in the forest clearing (*Lichtung)*, Vico had already located the original historical place in the *luci,* the forest that has been cleared for the first fields of golden grain. In both cases, Vico and Heidegger intended to link the emergence of human historicity with the imaginative activity of poetic consciousness.[9] Grassi concludes that ontology will then bear the indelible stamp of poetic insight. Consequently, we must expect that Vico and Heidegger focus on the significance of historical human reality; they do not share the rationalist's preoccupation with what is true. Indeed, the well-known ontological turn in Heidegger's work indicates that the ground of being can never be pigeonholed by discursive thought. Vico and Heidegger, then, are more compatible than Marx joined with either thinker, since the primordial opening onto reality is mediated by the poetic word and metaphor. However, when one considers the attention which Vico gives to the primary tropes of synecdoche and metonymy, an argument surfaces that more strongly links Vico (and consequently Heidegger) to Marx. That is, the explicit assumption of the body or parts of the body in the activity of metonymy and synecdoche clearly emphasizes the application of the physical self that is entailed in Marx's rendering of productive activity. Though a material product might not result from the application of the imaginative in Vico and Heidegger, the activity of poetic wisdom in the face of the ground of Being is no less productive, if the primordial poetic voice makes the human or civil world possible in the first place.

Following Grassi's comparative analysis of Heidegger's conception of the *Lichtung* and Vico's interpretation of the *luci,* one could formulate a position which sheds more light on the connection between the two thinkers. We might be cautioned in linking the primacy of the poetic in Heidegger and Vico, since Heidegger's claims for the poetic character of human existence or *Dasein* form a part of his call for the end of philosophy, and Vico's analysis of the poetic character of human apprehension refers to a so-called pre-ontological time in which the civil world first emerges. Are they not referring to separate temporal settings that must counsel restraint in linking the two thinkers? In both instances, however, each acknowledges a configuration in which human existence interfaces with totally other ontological forces that can only be fathomed to the extent to which the corporeal and spontaneous features of human existence are extended onto this ground. The power of personification, Vico acknowledged, tempers the frightful powers of nature as they are transformed into the familiar sphere of human personality. Grassi's assessment of Vico's *luci*, the space in which the ontological grounding of human existence occurred and Heidegger's *Lichtung,* a similar clearing in which the appearance of beings are made possible in the first place, deepens the affinity between the two. In both cases, though these clearings have the status of primordial ontological stages that are equivalent to the ground of Being, they had to be fashioned by human intervention. This can only strengthen the case that both Vico and Heidegger's conception of *poesis* is analogous to Marx's interpretation of *praxis* or free, practical, creative activity which entails the corporeal imposition of human productive forces onto reality.

Vico relates the story of the original folk who created this clearing to meet their material needs, and we have to assume that Heidegger's *Lichtung* is the result of humans cutting a path, *Holzweg*, into the dark forest to obtain wood for shelter and fuel. The real force of these metaphors urges us to question the conventional criteria for truth, since they figuratively point to the character of truth as emergence and appearance, after an area has been unconcealed. They equally serve to establish the more profound onto-poetic character of Vico's and Heidegger's interpretation of the primordial seat of existence. To quibble over the various interpretations of activity and action embodied in Vico, Heidegger, and Marx is to obfuscate the fact that the original ontological circumstances were made possible through human intervention. Marx's reflection in his third thesis on Feuerbach encourages the form of self-consciousness that is not only aware that humans are products of circumstances, but, even more powerfully, that humans are indeed responsible for those circumstances. Vico and Heidegger implicitly concur with Marx since the primordial clearing they each postulate is the setting for ontological circumstances that could only have been possible by the force of human labor. The mytho-poetic posture of human historicity that creates the circumstances of that historicity continues to fashion human historicity but, above all, this poetic disposition created the original setting that enabled the dawn of human historicity.

Heidegger's challenge to the Western philosophical tradition is more notorious than Vico's. Coupled with the primacy of feelings and moods which are predispositions for understanding in *Being and Time*, Heidegger's linguistic inventions and the appeal to the mytho-poetic give his work a disturbingly irrational ring. His detractors have sensed that his oblique critique of the tradition is no excuse for the lack of apparently rigorous argument. Heidegger never intended that irrationalism supplant philosophical rigor, but, in his critique of the conceptually bound ontological tradition of the West, he did recommend that thinking could in fact become more rigorous.[10] Like Vico before him, the dependence of philosophy on conceptually mediated thinking established that experience of reality had become fixated on a subject/object dichotomy that passed over the vital practical interchange that historically grounds thought. To conceive of Being as historical or as having a history, as Heidegger did, meant that his challenge to the tradition required an acknowledgment of human historicity that his critics have viewed as a distortion of traditional philosophy. However, his efforts to purge philosophy from its metaphysical conceits in his later philosophy are not the wholehearted deviation from the transcendental philosophy that occupies *Being and Time*. The recommendation, in his later work, that humanity become poetically poised, if it wishes to encounter Being, is adumbrated in the analytic of *Dasein* in *Being and Time*.

Some critics find Heidegger's penchant for new vocabularies and peculiar neologisms especially annoying. His inventions, to some, sound irrational, but we should expect a real challenge to the system to come with the introduction of new ways of thinking. Heidegger's appeal to a way of thinking more rigorous than the conceptual was not an appeal to irrationalism, but a call to reach layers of existence that had been obscured by philosophy's ambition to be scientific. Heidegger's appeal to a more fundamental and rigorous form of thinking beyond the conceptual [11] sought the help of a more phenomenological way of seeing. Getting beyond the conceptual required an articulation of the ecstatic dwelling of human existence in *Being and Time*, to which he returned in his later meditation that "poetically we dwell." The traditional representationalism, which is the heritage of Western thought, according to Heidegger, does not acknowledge the proximity of human existence to Being. Heidegger metaphorically illustrates this in his consideration that *Dasein* is the "shepherd of Being." To exist in the openness that makes the subject and object possible in the first place requires that philosophy pursue a questioning that actually experiences.[12] This form of phenomenological seeing corrects the representational thought of traditional philosophy, but it comes with the drawback, Heidegger acknowledged, of leaving us all too frequently with nothing to say. We would expect this result, if the poetic stillness that is required to reveal existence's proximity to Being discovers that language is not merely a tool that is wielded in the service of practical pursuits, but the essential structure within which we dwell. The explicit enmeshment of this ontological relationship is clarified in the historiographical influences of Heidegger's work (which is treated in the body of this work) and is especially reflected in Dilthey's interpretation of psychic life as the continual experience of unity. (Dilthey introduced the expression *Erlebnis,* in contradistinction to *Erfahrung,* to describe the essential unity of all life in which reality is imaginatively apprehended.) Living inside language, as Heidegger claimed, negates a subject/object dichotomy through which an experience of reality is subsequently labeled. As with Dilthey's conception of the acquired psychic nexus, Heidegger's analytic of *Dasein* dictates that the experience of life is not pieced together from elemental parts, but always comes as an historical totality. (Heidegger's desire for totality and his penchant for neologisms, which some critics say render his work indecipherable, is associated with the *Wandervogel* youth movement after WW I. See Peter Gay's *Weimar Culture* (1961), pp. 77–86 for such an argument. Gay's account places Heidegger's writing in the restricted historical context of Weimar Germany and does not connect it with the broader philosophical tradition. Heidegger's work reaches beyond the neo-romanticism of the twenties, though he was an inspirational voice for those looking for salvation in poetry and poetic philosophy.)

When Heidegger referred to language as the house of Being in his later work, he also advocated a return to the homeland. This aspect of his work has been cited as a vestige of his flirtation with nationalism. In the context of his later work, however, his appeal to a more rigorous form of thinking, which precedes representational thought, should not be overlooked. The rigor comes with the strenuous trip to the historical homeland that is nearness to Being.[13] The clue to this characterization of language, I believe, lies in understanding what can possibly be meant by language actually experiencing; that is, achieving something beyond the scope of mere reference and labeling, and understanding the difference between language construed as a tool and language interpreted as a dwelling place. What is meant by Heidegger's assessment that the essence of language ought to be conceived as the "home of man's essence"[14] implies that the differentiation that arises from speech conceals the primordial ground of language which comes originally with experience. Vico clearly intended that the primordial linguistic apprehension of reality was not distinct from reality, but arose simultaneously with the experience. While the way in which Heidegger called for a more rigorous sort of thinking beyond conceptualization was idiosyncratic he, like Vico, recommended a return to the pre-reflective origins or *heimat* of language, to the "dawn of thinking."[15] For Heidegger, the rich imagery of homeland that he found in Hölderlin's poetry only dramatized the homelessness of modern existence, which he conceived as a forgetfulness of the history of Being perpetrated by technocratic concern for material entities. A return to the homeland, on these terms, was a poetic reiteration of the historical stretching along of *Dasein,* which immediately granted a nearness to Being. The picture that Heidegger painted of the ecstatic temporal nature of *Dasein* is really that transcendence which grants *Dasein* the necessary structure to be close to Being. The Care structure enunciated in *Dasei*n is later echoed in Heidegger's claim that "man is the shepherd of Being." [16] Heidegger clearly explicates the derivation of the shepherd imagery from his articulation of *Dasein* in his "Letter on Humanism." [17] The imagery of the solicitous shepherd poignantly reflects Heidegger's appeal to a form of thinking and questioning that actually experiences. The experience that is synonymous with thinking is the thrownness of *Dasein* that makes us a caring neighbor for Being. The thinking that experiences, then, is the experience of care for Being.

Marx's efforts to restore humanity to its historical foundation were more obvious. His refutation of Hegelianism displaced the idealistic tenet that rational consciousness *per se* is the mediating force that overcomes the persistent alienation of humanity. Marx's continuation of Feuerbach's efforts to humanize philosophy is richly expressed in his conception of *praxis* or work. In extolling the liberating effects of free practical activity, Marx also advocated a return to origins. This interpretation of origins entailed the abolition of *Selbstentfremdung* and the restoration of the true social status of the human character. Marx's transformational critique of Hegelianism was a reiteration of Aristotle's rejection of Platonic metaphysics; both critiques acknowledge the practical basis for interpretation and understanding.

Marx's critique was influenced by Feuerbach, but his appeal to reform mystical consciousness appears earlier in his work, before he so violently rejected Hegel. As Shlomo Avineri points out, Hegelianism contains the seeds for an immanent critique of which Marx was fully aware.[18] It is in Marx's work that a theory of work (*praxis*) becomes the foundation for a philosophical humanism. Simply put, when *praxis* displaces mystical consciousness (the religious consciousness that places the burden of resolving this-worldly conflicts on an extra-historical agency) humanity becomes open to the radical way in which it is responsible for its own development. Most critiques of Hegelianism stress that Hegel's interpretation of history obscures real life, primarily in its reduction of reality to a purely rational compass. Marx's conception of *praxis,* especially, transformed Hegel's objects of consciousness into real tools with which to change history. Marx's conception of *praxis* echoes Heidegger's concepts of *Vorhandenheit* and *Zuhandenheit*, which are the qualities of ready-to-hand and before-the-hand which typify objects which exist within the historical domain of *Dasein,* and which are framed by the greater existential landscape of being-in-the-world. Likewise, Marx's conception of *praxis* also restores humanity to that social nexus in which the individual not only faces the actuality of an object beyond consciousness but realizes the interdependence (mutuality) of human relations that is concealed in the form of alienated labor, which is merely tacked onto existence by arbitrary economic structures. *Praxis* restores human existence to its originating function of transforming reality, because it acknowledges the instinctual drive to create that precedes the division of labor imposed by arbitrary social structures. Heidegger's analysis of *Dasei*n also restores humanity to its original totality in which the structures that define human existence are appropriated in the form of possibility; and not to be understood as phenomena in sequence that are tacked on as they are mathematically tallied. Both Marx and Heidegger equally address the explicit responsibility that comes with human existence *vis-à-vis* historical self-development. All three philosophers, Vico, Marx and Heidegger, then, acknowledge that reality is the product of creative human existence; a reality that is continually shaped by the historical subject and regarded as true insofar as it has received the transformative imprint of historical existence. Any truth claims beyond the purview of creative, historical existence are academic; they require proof of a non-historical vantage point on the part of an historically seated individual or community. The Cartesian legacy, however, sanctioned the reduction of human existence to unadulterated cogitating functions. Subsequently, this mandate led to the privilege of methods that dismiss the corporeal and the praxical window onto reality. In his search for inviolable truths, Descartes turned his back on traditional disciplines whose approach to reality begins with the choices made through human circumspection. Disciplines like law, rhetoric and even ethics, which formed the backbone of the humanistic curriculum, rested on shaky foundations, since they so readily assumed the existence of a world within which choice and action arise simultaneously with existence itself. Descartes' resolve to call into question all that he previously presumed true entailed professing that he had no body and that there was no world within which he lived. (*Discourse 4*, p. 54) Questioning the validity of a world and his body, though, was not enough to eliminate the presence of his own existence. The delusion of this meditation left Descartes with a disembodied, cogitating self and blessed the philosophical schools that embraced his method with an analytical voice assured of its own meditations but without a real world within which to resonate. The wedge that Descartes drove between the mind and the body has left some philosophical quarters without a grounding in actuality. If the intellectual purification that Descartes sanctioned is necessary in order to systematically discriminate, to exhaustively classify, to abstract regularities, and perform all the other functions with which thinking is occupied, then it is also necessary to build that material world required for mental substance to do its bidding. The unadulterated thinking substance that needs no place and depends on nothing material is an "I" without history, without heritage, without the instantaneous existential interchange between mind and matter, but it is the voice that speaks with the precision of the geometer's demonstration. Consequently, it gives expression to a vitality that can be found in the relationship of vertical angles formed from two intersecting straight lines. The understanding that can be garnered from the intuitive precision of geometry does not automatically compel us to apply this knowledge, since the Cartesian mind is detached from a world that it has yet to conclusively verify as actual, and would only verify through faith in a benign, non-deceiving God. In fact, there is no compulsion whatsoever to posit a world as long as the mind is content to accumulate clear and distinct ideas. Moreover, Vico made the point, over and over again, that geometry, despite its intrinsic logical structure, was another human invention.

Cartesian detachment was immediately challenged in Vico's poetic metaphysics, which we have located in both Marx and Heidegger. All three deny an autonomous mind operating without the material sustenance from, and attachment to, the historical world. They have all doubted the Cartesian detachment with an appeal to the hermeneutic impulse that accompanies all human understanding: that there is self-discovery whenever understanding and explanation is uncovered, and that there is no real understanding without the impetus to action. Heidegger, especially, outlines the manner in which existence itself is comported toward understanding, and, even more poignantly, how all understanding is joined with a fore-having or projection into the future, without which understanding would not have its source in human existence. The circumspective vision of human existence, explicit in Heidegger's analytic, and passed on to Gadamer's hermeneutics, entails a temporal foundation that is present in the form of historical influence, in the form of factical presence, and simultaneously projected as future possibilities. This temporal footing implies a value-related perspective that is skeptical of any underlying universal principle that is derived solely on the basis of unadulterated cognition. The vacuum that Descartes created in his philosophy denies the contiguity of mind and nature, which is expressed in the hermeneutic clarification that human existence is temporally poised onto its own historical reality. Once Descartes pretended that he was without body and place, he reduced his existence to the logical elegance of the geometer's compass and rejected the methodological premise that inquiry should be guided from the matter of which it treats, (assuming that an object of inquiry already speaks with its own kind of existence). Like the definition of the point in geometry, Descartes reduced human existence to the peculiar status of having representation without having place. In providing a vantage point from which to think, Descartes eliminated a world within which to think; he would require faith in a benign, non-deceiving God to resurrect the material world he had denied in order to exalt the primacy of his mental self.

The onto-temporal structure outlined by Heidegger, and received by Gadamer, is a totalizing process in which an already concretized heritage is aligned with the contemporary fact of human existence, which in turn is poised toward future possibilities. As such, human existence is not historical in the sense of being composed of events that have happened in the past; this would mean that human existence is formed of sequential events that are devoid of hermeneutical force. The ontological outcome of Heidegger's existential analytic indicates that human existence is equally joined to the past, present and future in a circumspective dynamic in which understanding is merged with the appropriation of possibility. It was Gadamer that exposed the ethical ramifications of Heidegger's analysis of human historicity, but the emphasis on the relationship between the *Geisteswissenschaften* and the ethical disciplines had already been anticipated in Ernst Troeltsch's monumental work *Der Historismus und seine Probleme* (1922). Troeltsch argued that the customary expression *Geisteswissenschaften* should be the general designation for the historical-ethical sciences, systematic, comparative sociology and the disciplines which interpret history.[19] Gadamer's position is closer to Troeltsch's because of the explicit applicative aspect of his hermeneutics, but Heidegger was careful to avoid any normative claims in his ontological rendering of history. Heidegger, however, was deeply impressed by the neo-Kantians in the formative years of his career, and Gadamer's distillation of the ethical impulse in philosophical hermeneutics discloses this early influence. Troeltsch also relied upon the ancient Greek interpretation of ethics to characterize the *praxis* explicit in his material philosophy of history.[20] Though he did not flesh out a material philosophy of history, Troeltsch referred to it in terms that make him a precursor to Gadamer. The praxical drama of history, for Troeltsch, meant that an individual, in a given contemporary formation, was inevitably "a task and a must at the same time," since all historical movement is rooted in a continuation of an already established concretized general store; what Gadamer might simply call tradition. Gadamer identified the weight of tradition as a positive prejudice and, like Troeltsch, distinguished any investigation of history as active historical life, which is always a formation of itself. [21] Gadamer, Troeltsch, and even Heidegger, intended that history construed as the investigation of singular events could only follow from histo-ontological foundations, which always assumes the problem of discerning the future development of historical life "out of historically understood presence" (DH, p. 79).

Troeltsch's project was directed at the philosophical conquest of historicism. To this end, Troeltsch, like Gadamer after him, demonstrated the positive value of the deep, penetrating, mutual relations of the judgments of the past, the present and future possibilities to develop out of these judgments, and how this conceptual chain of historical influence is praxically and normatively conditioned. This conditioning process provides both content and direction that makes an advance into the future a necessary outcome of historical interpretation. Critical of what he called the empirical quietism typical of conventional historical interpretation, Troeltsch's ambition to philosophically conquer historicism was guided by the premise that all contemplation originates in value connections that are generated out of the historical horizon which conditions our existence (DH, p. 116). The Archimedean position that Troeltsch searched for was not a mere point but an encircling horizon in which all contemplation immediately creates connections of value and meaning that press upon the vantage point of interpretation (DH, pp. 115–116). Though interpretation may derive from the "own situation" of the historical observer, its contents and meaning surpass the values immanent to the present moment. The autonomy of immanent values can only be fleeting; it is suspended, as Troeltsch wrote, between past, present and future meaning which contains the praxical energy that directs all interpretation (DH, p.117). What is clear in Troeltsch's interpretation of history, that is obscured in Heidegger's analysis, is the starting point in empirical history. With Troeltsch and other neo-Kantians at the time, the significance of the practice of the historian disclosed the ontological understanding implicit in the existential elements of the historical point of view. Heidegger had relied on Troeltsch's work in the very beginning of his academic career, but there are only small traces of this influence by the writing of *Being and Time.* Their careers overlapped, and Heidegger was even considered for Troeltsch's chair (along with Ernst Cassirer) at the University of Berlin in the late 1920s. What Heidegger's work conceals, but which is revealed a generation later in Gadamer's work, is the vital role of the aesthetic and ethical dimensions of the historical point of view. The explicit hermeneutics of historical inquiry for Gadamer was highlighted in the aspect of application, which is tied to understanding in the same manner that the past is linked with the future. The origin of Gadamer's and Heidegger's hermeneutic outlook in the practice of the historical investigator, whatever form this may take, means that historical understanding is enveloped by both methodological practice and the applicative aspect of human historical understanding. The neo-Kantian influence on the early Heidegger is quite clear in 1915, when he presented a qualifying lecture to the faculty at Freiburg. This bias was still alive in his critique of Rankean methodology in the final paragraphs of *Being and Time.* These influences suggest the challenge to ahistorical, universal foundations and conditions, which are accentuated in postmodernist efforts to undercut traditional philosophical foundationalism and objectivism. This contemporary disposition was less self-consciously exercised by philosophically minded historians and historically minded philosophers whose methodological inquiries revealed the depths of the historical imagination.

# 1

# Four Moments in the History of Historical Reason: The Structure of Internal Time Consciousness

### *Augustine's Meditation*

Since Augustine's heady meditation on time in the *Confessions*, the subject of temporal existence has been central to speculation on the nature of history. Augustine's recognition of the temporal constraints on consciousness has never grown out-of-date. The moment we raise the question of time and temporal existence, it seems hopelessly out of reach. This insight is equally a part of the historian's difficulty in clarifying the structure of history. On the surface, history seems quite obvious to us. Can it be anymore than an account of the past as garnered from documentary evidence of one kind or another? Is it anymore complicated than the remembered past? To cast it in these simple terms, however, raises immediate difficulties. Whose memory is being recalled; what guarantee is there that sources of information are accurate? These questions might not sound compelling, but they in their simple way point to the epistemological intricacies of historical thinking and even the metaphysical foundation of all thinking. The later point is demonstrated by the ever remoteness of the historical event that might be under investigation. Provisionally, an historical analogue to Kant's *Ding an sich,* which is equally beyond the purview of the synthetic processes of the mind, comes into play as soon as one raises these familiar questions. The fact that sequential coherence can be derived from the experience of life and formed into a conceptualized structure, with both continuity and the exclusion of related but irrelevant facts, suggests the function of *a priori* structures which were introduced by neo-Kantian innovations in the nineteenth century.

The problem as outlined by Augustine over 1,500 years ago remains the central difficulty in historical thinking. The positive outcome of this puzzle, however, indicates that Augustine had articulated a vital concern in explicating the nature of historical reason: the revelation that consciousness is temporally structured as to produce coherent totalities out of the continuity of life. His clarification of temporal existence is found in the dialogue he initiated with God in the *Confessions*. To imagine that he was conversing with God meant that he had to imagine talking with a transcendental entity, out of time, eternal and essentially without a second. If God is without a second, the first matter on the list of creation had to be time. From Augustine's point of view, this has to be the most puzzling of God's creations, for it is the source of the most complex form of reflection.

To even talk of time at all becomes a convoluted enterprise. Augustine begins with the acknowledgement of the specious present, that the moment has an ambiguous status, since it does not have duration in the same way that we speak of a long time in the future, or a short time ago in the past. Duration, which strikes us as integral to any conception of time, is completely absent from the present moment. Where then is the vantage point of my thinking, or anyone's reflection, on any matter? The specious present, as William James also understood, can only be an abstraction. But, where does history come from; that is, the events that leave a record and find their way into books? Is it not created in the present? Augustine, prefiguring Heidegger's existential analytic, understood that the answers to such questions must lie in the internal structures which house time. That is, history may indeed begin with experience of the empirical world, but its conceptualization requires organization which does not derive from experience.

We might reply that it is quite simple to keep tabs on time. Can't we agree that on Monday I did such and such a thing, and this makes the event three days ago if today is Thursday. Well, if I can recall what I did on Monday, then isn't it somehow still with me, in memory, as a record on the calendar? If so, Monday is really not over in my mind. Augustine acknowledged this condition induced by time, and so his meditation prefigures the neo-Kantian methodological deliberations which outlined the way in which history is really a transformation of life, since it originates in the temporal laws of the mind. Augustine's meditation uncovered the ecstatic and continuous nature of temporal existence. That is to say, there is no clear demarcation of the past from the present and likewise the future. The facets of time are really a cooperative movement in their mutual realization. Well then, how can history be written if it is always happening along this ecstatic continuum? How do the demarcations between periods manifest themselves? This is just not a problem for historians, it is a problem with thought itself, since we customarily regard thinking as an activity in the present, though it might be dwell on the past, or anticipate things to come. Augustine imagined that any genuine description of the past, the subject matter of history, is not equivalent to *past facts*. What is required, as called for by Georg Simmel, an exponent of the neo-Kantianism at the end of the nineteenth century, is an activation of the constitutive structures of the mind. An epistemological wedge may always distance history from life, Simmel surmised, but like Augustine, he maintained a metaphysical compatibility, between history and life, which has its origins in the volitional forces of existence. As Augustine put it, the past is described correctly when the mind produces words based on memory-pictures of the historical facts (Book XI,18). Most importantly, the present, which is the temporal, volitional seat of consciousness, is the vantage point for activating memory. This implies that time is an extension or inflation of the present. This is why Augustine spoke of the facets of time in terms of a present of past things, a present of present things, and a present of future things (Book XI, 20). Edmund Husserl's investigations in his lectures on the structure of internal time-consciousness (these essays date from 1904 to 1914 or so where compiled and edited by Martin Heidegger in 1928) echo Augustine's insights and most likely provided impetus, as well as terms and concepts, for Heidegger's ruminations on the temporal character of *Dasein*. Husserl, as well as Bergson, established the temporal grounding of consciousness, and more importantly, drew attention to the type of continuity and extension which distinguishes human existence from entities that are spatially framed. Husserl provides a more rigorous analysis than Augustine, replete with the phenomenological construal of objects as self-presenting and memory (history) as the source of re-presentification of images which have been "shoved back" (*Zuruckgeschobenes*) in consciousness but still overlapping the present which is occupied with objects.

The awareness of periods of time seems to be commonplace. But when we think of the past it no longer exists. How, then, does history, which is the account of the past, become tenable? Augustine's meditation complicated this issue by thinking in terms of measuring time. The past, or the future, cannot be measured for obvious reasons, so time must be calculated while it is passing. Our awareness of time is an acquaintance that comes from within, since the shadow of the past, and the fantasy of the future are not concrete. These customary divisions of time appear less and less as divisions, if the past and future are linked to an imaginative apparatus. The sanity of Augustine's position is overwhelming; for, how is it possible to see what does not exist? Isn't the prospect of doing history or thinking in time threatened by not being able to perceive what no longer exists?

Certainly history is something that has a concrete effect on the structure of political society, the economy, etc., but is it all a chimera, an illusion, or at best a sociological construct that is passed down from one generation to the next in the way that Emile Durkheim, from his peculiar sociological point of view, conceived the notion of God? The other alternative, as suggested by Augustine, is to consider the possibility that the facets of time derive simultaneously from some internal refuge whose peculiar activity is to absorb the separate facets of time, one from the other and into the other in a movement, which Augustine imagined to be like the consciousness of God. Otherwise, how else can the past be reported upon and the future anticipated? And what better dialogue to have if one seeks the mystery of the three aspects of time in that continuum called history.

Calculating the factual reality of the past and the future might be beyond the purview of human consciousness, but they are conceivable categories of some kind, when viewed from the present. But the present doesn't exist as a concrete vantage point, since it is so ephemeral. Augustine uncovered the dilemma of the historian, when he said that a description of the past does not constitute a fact, but is merely words based on memory. Without facts what kind of science can history be? Perhaps neither an art, nor a science, but an autonomous form of apprehension.[1] A form of apprehension, in fact, that describes human consciousness.

The theoretical position enunciated by the neo-Kantians, Simmel, Windelband and Rickert, and other exponents of the revival of nineteenth-century Kantianism, declared the independence of historical conceptualization and representation when they confirmed lived-reality as the object of historical inquiry, rather than the pursuit of conceptual validation. The conceptualization of lived-reality as the object and the source of writing history recalls Augustine's edifying meditation on time.

Carl Becker, in his illuminating address to the American Historical Society in 1931, suggested that the activity of history is endemic to us all. The ubiquity of historical practice is easily discerned in the procedure of tying a string around one's finger to alert us to a deed to be done, which can only be marked by a deed in the past.[2] The future, it appears, is inextricably tied to the past and the present. This does not imply prophecy, since there is only the formal suggestion of an anticipated act in the sign of the string around the finger, not the future-as-fact. Augustine's poetic example of the rising of the sun in paragraph 18, Book XI of the *Confessions* provides us with a universal example of viewing the future on the basis of a present sign. But then again, shouldn't we allow for the possibility that someday the sun will not run its regular daily course? It remains that although the future is undeniably intangible, it is a formal principle that lies embedded in the present. Why else would I tie that little string around my finger, it must be a sign that anticipates a future prospect. This activity, we can now correctly say from the vantage point of the explosion of historical thinking in the nineteenth century, is indicative of a critique of historical reason. The rationale for this judgment is the underlying presupposition, on the part of Augustine, that he is analyzing a universal cognitive capacity, much in the same way Kant outlined in his *Critique of Pure Reason*. The essential difference between these two considerations of a common we (the "I think" behind all experience) is the now customary division between the historical or civil world and the natural realm, which Kant did not make, but which was established in the wake of the nineteenth-century Kantian revival. On the one hand, the common we relegated to the natural world is concerned with the perception and discovery of sense objects and their relations, while the we of the historical world is concerned with the production of artifacts and their matrix of application. Clearly, Kant's critique concerns the how of knowledge and Augustine's and the historical critiques, which followed Kant, concern the "why for" of knowledge. Application evolved as the hallmark of the historical critique. Therefore, the limits of historical reason are only constrained by the productive imagination; they are not strictly reducible to inner conceptual categories. The eschatological vision, which launched Augustine's philosophy of history, corroborates the productivity of the historical imagination, since it was focused on the future. There is obviously an extra-transcendent aspect to Augustine's history, the Incarnation is the model transcendent feature, but this form of transcendence was supplanted by a metaphysical subjectivity with the emergence of Romanticism in the nineteenth century. Nonetheless, Augustine's early analysis of time-consciousness is as modern as the bifurcation of the historical and natural worlds, it is as relevant as the contemporary discussions concerning methodological differences in the disciplines, and it is as fruitful as the Marxian transformational critique. It has provided the presuppositional framework for a critique of historical reason. Augustine's acute understanding of the temporal nature of human consciousness established that time is not a mere imaginative construction, but the internal structure that measures practical activity. As the fount of the historical world, this temporal structure guarantees unceasing creations, and so it must be differentiated from the routine realm of nature and its regularities. Augustine established the verity of time as an inner continuity that can only be measured as it comes from the future, passes through the present and then recedes into the past (Book. XI, 21). The uncanny consequence of this movement, then, is that time emerges from what is not, passes through what has no duration, and then quickly subsides into non-existence again. In the midst of this peculiar continuity, though, practical life emerges. Therefore, from the inner space of this fleeting duration, history, culture, and the civil world comes forth. Augustine's perplexity can only be our perplexity in unravelling the mystery of time-consciousness.

Before the discussion over the distinction between the natural and human sciences began in earnest, Augustine disclosed that the power of time lies in its capacity to measure the internal extension of the mind. Time cannot be explicated in regard to the movement of bodies, since time persists without obvious, external movement. If it is independent from the movement of a body, then its extension must lie elsewhere. The practice of measuring the passage of time consists of calculating a duration that comes from non-existence, a not-yet (the future), and recedes into non-existence from a position that has no extension (the present). This passing interval, constituting a completion that is also a chimera, Augustine concluded, is a calculation of what is fixed in memory. This thorough empirical assessment of time resolves that the movement that is measured is a subjective impression. There is nothing objective to measure, only a residue of the moving temporal fantasy. Time, then, is a measurement of meaning that arises from the impression an experience creates. This must be the source of the various ways in which periods of time appear to be either more expansive or more restricted than they actually are. How time flies when we are enjoying ourselves, how prolonged a period becomes when we dread an experience.

Augustine can only be anachronistically credited with the conception of the phenomenon known as *Erlebnis,* so crucial to life philosophy in Germany, but the temporal organization of the mind he outlines is no less indicative of the continuity of the psychic set forth by Henri Bergson, and those like Georg Simmel and Wilhelm Dilthey, who helped introduce the conceptualization of experience as the embodiment of productive meaning in contrast to the objectivizing and spatializing concept of time in positivistic thought.

If time really resolves into an impression, then the matter of historical reckoning must begin with an after-image of sorts. History, broadly construed as the memory of past events, must also be regarded as an after-image, in which case it too must emanate from a subjective vantage point that is prohibited from ever gazing upon occurrences themselves (Hegel's conception of Original History notwithstanding). Furthermore, if we conceive historical events as the realization of motives and the internal structure of intention on the part of that entity we call human existence, then history-as-event is generated from the anticipation of what is not. Logically, this conception of history must mean that it is a creation out of nothing, since the future has the status of not being yet. Augustine's conclusion that nothing objective is measured when we calculate the passage of time will have certain consequences in the analysis of history as a discipline and as a manner of accounting. A definite analogy suggests itself, if a thing-in-itself is not the focus of measurement in the context of time. We might just as easily question the objective status of the historical object which is presumably the center of attention in the pursuit of history. We needn't argue very long to see the parallel in history. Where is the object in history in any historical account? Is it bounded in any way, does it form borders that distinguish it from other objects of its kind? If the historical object fills the "space" that is described in Augustine's treatment of time, then we can expect it to have the potential to indefinitely extend itself.

In a modern interpretation of history, as offered by Hans Georg Gadamer and the historical school he hails from, we can readily see the consequences of Augustine's analysis of time. For example, Augustine's meditation is evidenced in Gadamer's concept of *Wirkungsgeschichte*, since it too articulates the tendency of the past to initiate effects that will someday instigate and color historical inquiry. How this is so will be clarified in what is to follow, but the merit of Augustine's thought already suggests that historical investigation, and the measurement of time, is always a calculation of an effect, or arises from an effect which has persisted through time. If history, as an account, or the calculation of time, is always in response to an impression or the effects of the past, then time-in-itself, or history-in-itself, is not a thing but a calculation of an internal continuity that resides in the memory of the individual or culture. This must be the most remarkable and relevant conclusion for any understanding of history as an account or explanation of events. The internal continuity that makes up time, and history, provides us with the most fundamental component of a critique of historical reason. That is, historical thinking can never gain knowledge of a thing-in-itself. Such knowledge is out of reach not by any shortcoming of the mind, but by virtue of the self-mediating component of historical thinking; that is, history is self-constituting.

Insofar as history apprehends totalities, it must pursue them through an effective grasp of things to come. The context of historical experience, then, must be an encircling movement that intelligibly completes itself from within a view of totality. This encircling movement is addressed in Dilthey's concept of the "psychic structure of experience," which is really a process of continual acquisition. It is animated from within experience and thus cannot have the status of an object that stands over and against consciousness; it is phenomenologically inherent to awareness. Construed in this way, history is the apprehension of psychic effects whose objectivity is always linked to a projected totality. Hegel's final conclusion that philosophy can be a science and provide actual knowledge was heralded in Augustine's meditation, for as the future, which is not yet, is relegated to the past, the past will always increase in proportion to the future. Finally, Augustine argued, the future will be absorbed by the accumulation of the past and then and only then will a whole exist. This totality, according to Hegel, comes too late for philosophy to understand the world, but for Augustine there appears to be a way of calculating completion that has yet to be realized. In this respect Augustine's meditation resembles Heidegger's existential analytic in which the totality of *Dasein* is projected through a process of resoluteness.

The extension resulting from the facets of time collapsing into one another is really the mind being attentive to its three corresponding functions: expectation, attention and memory. We can certainly deny the factual reality of the three facets of time, but their presence is acknowledged as a form of historical apperception, which Augustine simply refers to as the faculty of attention. Since this extension is present while the mind is engaged in one of the three functions laid out above, it must be regarded as foundational to experience. As the precondition for experience, then, this form of extension can only be the focus for further analysis of what makes history possible. Augustine offers a more existentially oriented interpretation of time than does Hegel, for the source of the extension and attention of the mind is subjective, it is not governed by the dictates of an objective mind hovering above one's individual existence, as decreed by Hegel's infamous *List der Vernunft.* The proof of this is the capability of the faculty of expectation to engage a totality before it is realized. This capability was later addressed by Heidegger in *Being and Time,* in which it is conceived as the fundamental temporal orientation of "being there" or *Dasein.* Heidegger's existential analytic, then, is a reiteration of Augustine's claim that life can be lived forward by virtue of the mind's peculiar ability to engage anticipated totalities. Likewise, we expect this future orientation to have its counterpart in relation to the past. That is, historical totalities that consist of events that have transpired can be equally conceived as unities. The catalyst in apprehending the future, as well as perceiving the past, must have its source in whatever imposes the concept of meaningful totalities onto consciousness. Both Augustine and Heidegger would conclude that whatever the catalyst, which generates totalities in respect to historical apprehension, it must equally apply to one's own existence. Heidegger's philosophy, which was grafted on to the radical critique of religion instigated by Nietzsche and Marx, has eliminated the mysterious interlocution with the God of Christian eschatology. Nonetheless, it maintains the integration of historical totality elucidated by Augustine. Before Heidegger's interpretation of history is addressed, though, we must see how the critique of historical reason was fashioned by two other moments that constitute its development. The two other moments that stand out as expressions of a critique of historical reason can no longer be viewed as autonomous expressions. Marx's interpretation of history, that is, cannot be seen as independent from Hegel's philosophical history, and Dilthey's innovations, thanks to recent scholarship, can not be seen as separate from Vico's revolutionary principles. The trends in scholarship dictate that the Hegel/Marx movement and the Vico/Dilthey constellation are mutually encompassing. This is clear by Marx's own admissions and recent Vichian scholarship, beginning with Benedetto Croce, imposes the Dilthey/Vico constellation on us. I have elected, though, to look to Vico, whose philosophy is more obscure to twentieth-century philosophy, since Vico argued for the tenability of both a historical logic and a metaphysics free of a rational architectonic.

## Vico's Historical Epistemology

Vico's resurrection in recent years has led to a new understanding of his place in the history of philosophy. It seems that his legacy, for a while, had been dictated by his difficult and obscure life. His recovery has been largely due to Croce's assessment of his understanding of the historical world and most recently, in the United States, from Giorgio Tagliacozzo's dedication to the dissemination of Vichian scholarship.[3] Croce's judgment would have us believe that Vico was the first to acknowledge the consequences of the epistemological gap between the historical and natural world, and therefore the fundamental difference between the human studies and the natural sciences.

Vico had been imported into nineteenth-century Germany, as Isaiah Berlin recounts, but his importation came after both Herder's and Goethe's original views on the historical world.[4] Herder can take much credit for the first expression of romantic historicism, and Goethe, according to Meinecke, can be regarded as the epitome of the historicist mentality. But Vico cannot be regarded as an influence on the historicist movement that took hold independently in Germany.

Several of Vico's theses are of paramount importance in sketching the development of the critique of historical reason. They illuminate the two basic principles that Dilthey in his own way has introduced: (1) that there is a fundamental distinction between the natural and historical worlds and (2) the logical corollary that a method of understanding, distinct from the methods peculiar to the natural sciences, is required in order to make the historical world intelligible. Vico's major insights affirmed that the historical or civil world was of humanity's making, and therefore the understanding of this world must lie within the structures or modifications of the human mind.

The assertion that the historical world resolves itself into the modifications of the human mind also led Vico to a rudimentary formulation of a principle of understanding and explanation which parallels Dilthey's concept of *Verstehen.* Vico postulated his version of sympathetic understanding, however, within the context of an ideal eternal history, while Dilthey's aspiration for a critique of historical reason was in direct response to Kant's understanding of experience and knowledge as a transcendental synthesis. Nonetheless, their philosophical ambitions reflect a very simple notion: that humanity creates the historical world and consequently can only be understood as a product of the historical world. This simple premise has been the focus of much Vichian scholarship, and it is now customarily regarded as the core principle establishing the bifurcation between the historical and natural worlds. Vico's articulation of this principle in paragraph 331 of the *New Science* has been expanded to both include an epistemology, as it necessitates a form of knowing different from that in the natural sciences, and a form of historical criticism or understanding, which also must be distinguished from the methodology of the natural sciences.

Vico's understanding of the separate form of knowledge applicable to the historical world was established within the context of an ideal history, which was not unlike the earliest form of the ages of man in Hesiod or the more modern analysis set forth by August Comte. The three stages of humanity in Vico's scheme, as in Comte's scheme, correspond to respective mental attitudes or worldviews. Vico's three stages sound very much like Comte's in fact. Out of the three, the time of the gods, the time of heroes, and the time of man, the first stage was occupied with the most interesting form of expression. Vico's tendency was to measure the mental disposition of each stage in terms of the development of language use. The structure of language, he argued in a very modern way, provides us with insight into the deeds of people.

The original linguistic structuring of the world was predicated on a poetic logic that tended to ascribe characteristics of the body to the world, to the extent that actual forces in nature could be represented by gods personifying certain attributes. Vico, needless to say, was deeply moved by the worldview of Homer and Greek mythology in which this poetic structuring of the world is evidenced. The poetic logic of the earliest peoples, according to Vico, antedated the emergence of conceptual apprehension in which the things of the world are absorbed by mental categories representing genera, but it too transforms the world into types. Contrary to the categorial absorption of things, however, one's sensuous being is extended into the world in an effort to achieve understanding. This imaginative process delivers the degree of acquaintance necessary for knowledge, but it involves the transformation of things into the personally sensuous activity of the self.[5]

This poetic metamorphosis of the world at the earliest philological stage created an orientation to the world, but the constructed domain of familiarity was built on the foundation of concrete relations which Vico would classify as poetic or imaginative universals. Such a philological orientation will make one tremble in the face of lighting and thunder, which comes from within the *topos* of the gods, rather than lead to an analytical understanding of certain forces of nature. Still, the orientation to the topical resembles an understanding of the typical. It is an understanding that conforms to the sensuous apprehension and transformation of the things in the world, in contrast to the conceptual apparatus of sensibility and understanding. This poetic apprehension and transformation of the world has its source in the original tendency to create names for things on the basis of familiarity with one's own body and productive activity. Language is still peppered with these original connections of the body with objects and relations in the world, and so we can conclude, along with Vico, that historical development and the stages of philological apprehension are never made obsolete. This conclusion is the basis of any effort to adduce a framework for historical criticism and historical understanding, since it implies that there is no philological world view that is alien to our own. At any present stage of historical awareness, we are subject to the internal continuity captured in the evolutionary dialectic of ontogeny recapitulating phylogeny. Vico's rendering of this historical principle is captured in his understanding of the *sensus communis*, that linguistic and interpretive orientation to the world that comes with the force of tradition and the heritage of custom.

The validity of poetic or imaginative universals, then, is established in the movement from particulars to generic types or universals. It is the unmediated presence of one's own existence with which the process of creating poetic genera begins. Vico's interpretation of the original philological apprehension of the world is surprisingly modern, since it is reminiscent of Marx's critique of Hegel and the consequent inversion of the Hegelian cosmic drama. This inversion, or transformational critique, also recognized the validity of beginning with the sensuous apprehension of one's own activity, rather than to assume the canonization of the subsisting, as prescribed in Hegel's logic. Marx's contribution to the critique of historical reason will be addressed in the next section, but we should acknowledge the correspondence between the imaginative universal, as a product of human activity, and Marx's intention that the conditions for the creation of social relations and the overall cultural matrix are always inventions of practical human activity. Construed otherwise, humanity will only persist in a state of what Marx called *Selbstentfremdung* or self-estrangement. As the creator and product of history, human existence defines the conditions that transpose knowledge into practical capacity.

Understanding historical phenomena, then, requires a penetration of the rules that generate a given philological apprehension. The autonomy of a given philological model, for Vico, indicated that language produces thought. Consequently, comprehending a certain model will provide insight into the way people thought at a given stage of history. This simple principle promised what Vico called a "metaphysical art of criticism," which is ultimately proved by the "common sense of the nations" (*sensus communis*). [6] The common sense of the nations, though, rests on an even more fundamental foundation–an ideal eternal history decisively established by divine providence. This sounds remotely archaic, but neither Hegel nor Marx had completely dismissed this eschatological orientation by the nineteenth century. The providential course of history was a facet of Vico's Catholicism, and a necessary postulate to explain human development from *stultis* to *sapiens*. While some scholars avoid the sentimentality imbued in Vico's concept of providence, Mark Lilla's comprehensive work on Vico, *The Making of an Anti-Modern* (1993) finds the thread of the divine *conatus* (will) throughout the body of Vico's work.

This unquestionable principle, found in paragraph 331 of the *New Science*, does not lose its vitality in the context of this eschatological postulate. The world of nations (the historical world) is most certainly a product of human activity, and so understanding the origins of this world will center on the modifications of the mind, as Vico put it. This principle has been traced to Renaissance sources, but we might even look to the ancient world for the prototype of this theory. Was it not Socrates, in his dogged pursuit of wisdom, who invariably focussed on the ability of the craftsman to satisfactorily provide explanations of processes and phenomena on the basis of a body of knowledge, *episteme,* underpinning a practical activity or *techne.* Socrates's repeated use of this craft analogy was designed to clarify the nature of the good life and how it might be secured, but it embodies the very same principle around which Vico's historical epistemology revolves. The Vichian principle of *verum et factum convertuntur*, the convertibility of knowledge (truth) and the reality of a product is contained in Socrates's constant reflection that the craftsperson is the incarnation of the reciprocity of knowledge and creation. The origin of the historical world, then, if it is the creation of humanity, must lie in the permutations of the human mind. Socrates's use of the craft analogy and Vico's principle verify that when someone creates something, they can also transmit these creations in the form of systematic prescriptions that can be subsequently followed by others. Vico acknowledges the reciprocity of creation and narration,[7] and clearly Socrates stresses this correspondence in the integrity of *techne* and *arete*, the end or functional excellence to which the craft is dedicated. Practical activity is always dedicated to the particular goal which fulfills the nature of the craft, as set forth in Book I of the *Republic.* Vico's principle, like Socrates's use of the craft analogy, was committed to a much broader hermeneutic tool. Socrates's model proposed an imaginative way to envision the elements required for the attainment of the good life, and Vico's suggestion that there were systematic and uniform philological orientations established common sense as the foundation for a synchronic art of criticism.[8]

The realization that common sense is typical of all pre-reflective judgments led Vico to assert that it is the primary criterion of criticism which would finally lead to what he called a "mental dictionary," through which one could decipher the nations and diverse languages, as if from a glossary of a common mental language. This postulated lexicon did not prove that nations and states are derived from an original parent nation. On the contrary, Vico formulated the historicist principle that civilizations with similar customs and laws had separate origins.[9] Vico's positing of such a "mental dictionary" goes to the heart of the contemporary hermeneutic preoccupation. If, for argument's sake, we agree with Schleiermacher that hermeneutics or interpretation is relevant wherever misunderstanding is a possibility (or wherever there are words, the source of all misunderstanding according to the Little Prince), then we can see that Vico imagined the feasibility of a universal lexicon that could penetrate the most alien and bygone of epochs. The presupposition, here, is that translation is the model for hermeneutic reflection. That is, it is conceivable that there will always be an unfamiliar language whose comprehension will entail a reference to a third source for resolutions of meaning, that is, a dictionary. Vico did not provide us with an articulated program for the human studies, but his postulate of a "mental dictionary" that could decipher the uniformity of ideas arising among varied populations proposes a definite grounding for analyzing the degree of uniformity among diverse philological apprehensions. Such a common dictionary, although its construction requires a view of the limitations of cultural expressions that are conceivably still unrealized, would serve as the foundation for entry into a given historical context. The consequences of such a postulate presumes a metaphysics of culture or history whose presuppositions are only realized in relation to a particular historical context. As such, Vico can be credited not just with the postulate of a "metaphysical art of criticism" but with a metaphysics of culture that is predicated on the correspondence between a given philological apprehension and philosophical truth. Such an isomorphism recommends the union of the certainty of philology (historical facts) and philosophy which aims at the general. Philology possesses the certainty that is characteristic of historical facts, since it is the result of human choice, while philosophy holds out for the tenability of a general framework embracing the limitations of knowledge related to matters of human choice.[10] This was summed up in Vico's distinction between *certum*, which lies within the province of the human world, and *verum,* the domain of eternal truth.

Philosophy, then, orders all the manifestations of human choice, and history or philology is the space for the appearance of all the particulars. Hegel, for example, sacrificed the permutations that can occur, in reference to human choice, in an effort to make all conceivable historical manifestations of human will intelligible from the point of view of an ordering principle that straddles both history and the extra-historical realm of the Spirit. Vico regarded this tendency, on the part of philosophy, as a form of conceit that peremptorily dismisses certain viewpoints in its effort to tell us what we should do.[11] Consequently, philosophy can only be of interest to those who think they know what should be the case. Hegel exemplified the philosopher's conceit, since his dialectical scheme embodied the rationality of the subsisting, and overlooked the primacy of the sensuous mediation of human existence in the creation of the historical world, which, according to Vico, is grounded in the philological apprehension of reality.

The instance of poetic universals, especially, illustrates the empirical or historical grounding of philological apprehension, while at the same time guarding against the philosopher's conceit. As language, the imaginative universals do not attempt to reproduce reality as in prosaic, linear, descriptive speech, which, when relying on claims of an inherent atomistic correspondence between speech and reality, overlooks the given alienation between the internal mechanisms of philological apprehension and the external world of material extension. Vico, before Hegel, proposed that the alienation or opposition between mind and the external world is overcome in a process of the negation of the negation; that is, the transformation of the alienating world of matter into the familiar processes of corporeal human existence. Vico's version of the negation of the negation is not an unadulterated mental process, as with Hegel. It is a form of poetic thinking which entails a projection of the self into the things and relations of the world, to the end of transforming things and their relations into the familiar sphere of the body, with which we are naturally acquainted. This does not entail hanging a label on a thing or a relation, since synthesis is initiated from the vantage point of what is already corporeally familiar. Well, we might just ask if all such transformations are poetic in character, or if there is really any other alternative in achieving knowledge of the world. If the poetic apprehension of reality originates in particular contexts, that is, contexts that are always familiar because of the given acquaintance of my own being, then it might very well infiltrate all forms of historical apprehension.

Heidegger, after his thorough analysis of temporal existence, found philosophical refuge in the thought that humanity dwells poetically. He, too, as we will see, favors the origin of thought in what he identified as the immediacy of being-in-a-world (what we are calling sensual), but it is first to Marx that we should turn to see how Vico's suggestions for a metaphysics of the sensuous plays itself out in the furor of the emergence of historical thinking in the nineteenth century.

## Marx's Transformational Critique

Apparently there is only one brief reference to Vico in the body of Marx's work. The brief reference, however, signifies that Marx was familiar with Vico's fundamental insight into the historical dimension of human existence. Any influence from Vico is dubious, but the recognition of Vico's assessment that the historical world and its creations are more accessible to the understanding than natural history make Vico and Marx compatible philosophical types.[12] Moreover, Vico's repudiation of the Cartesian world view, and Marx's critique of Hegelianism, compel them both to question the validity of system building on the basis of a conceptually derived first principle. In the case of Vico, Descartes' ahistorical foundation of the certainty of doubting one's own existence led to an overemphasis on logocentrism and the consequent displacement of methods of inquiry and discourse founded on the assimilation of sensual schemata. Ernesto Grassi, whose work recognizes the affinity between Marx and Vico, has questioned the dualism of pathos and logos in the Western tradition. He has argued that the institutionalization of the anonymity of rational discourse neglects the essentially political nature of the community and human nature. Like Gadamer, who argues for the vitality of hermeneutics in *Truth and Method,* Grassi recommends the resuscitation of Aristotle's practical philosophy and the cultivation of the faculty of *phronesis*, that capacity which can function in particular situations without the benefit of theoretical knowledge of the general.[13] Marx would assume a similar position against Hegel, who also relied on the so-called supremacy of the conceptual in a systematic effort to corral the manifold of nature and history.

Marx and Vico, then, can be seen as exponents of what Grassi, and Vico himself, called topical philosophy, in contrast to critical philosophy. Roughly, these philosophical approaches correspond to the distinction between the human studies and the natural sciences. Vico, and Marx after him, identified the epistemological credibility of the more pathetic inquiry and enunciation of the human situation, that is known as topical philosophy. As the name suggests, this approach, in contrast to the construction of a rational science founded on the cognitive certainty of a first principle, aims at understanding the particular, which is synthesized in a sensuous assimilation of the historically concrete. The unadulterated rational approach, Vico thought, tended to exclude the kind of truths that are not generated through quantitative analysis, as in the case of mathematics or physics. Vico's inheritance of the humanistic tradition's epistemological orientation of knowledge *per caussus*, that is, the formula emphasizing the reciprocity of knowing and making and the tacit familiarity with origins, must be measured by the practical value implicit in all scientific inquiry. Marx's critique of Hegelianism rests on this very same principle; that is, the orientation of the science directed toward the elimination of the traditional status of society and the subsequent emergence of a social order claiming only a human status[14] must recognize the limits of scientific inquiry predicated on the primacy of the cognitive, rather than on practical value. This new humanistic science, then, must mean that its validity is identified with the underlying movement toward invention and not discovery.

How Marx's critique of Hegel took shape begins with Feuerbach's transformational critique of Hegel. This transformational critique is recognized in Marx's notorious saying that philosophers have hitherto interpreted the world, while the real point of philosophy is to change it (XI Theses on Feuerbach). Feuerbach had begun the work of understanding the role of the sensuous in contemplation, as Vico had earlier recognized in mytho-poetic consciousness, but, according to Marx, he had not fully appreciated thinking as integrally connected to activity. Hegel's philosophy met with considerable criticism in the re-vitalization of Kantianism in the late nineteenth century, but his cosmic architectonic was first conceived in earthly terms by Marx and the Left Hegelians. The division between the Left and Right Hegelians can be seen in Hegel's dictum: "What is rational is real; And what is real is rational."[15] The latter portion of this synoptic grasp of Hegel's systematic approach to reality expressed the world-view of the right Hegelians, while the sentiment expressed in the rational being equivalent to the real is more typical of Marx's position. Marx's inversion of Hegel's logic meant that humanity's accustomed self-estrangement could be overcome by the realization that a given cultural configuration and the relationships therein were actual historical creations, not the work of an extra-historical omniscience.

This resolution, of the religious essence into the human essence, circumvents the overarching determinism in Hegel's system and eliminates the need for providential explanations, since it confirms the awareness that though we appear to be thrown into given circumstances, they are cultural products and thus subject to modification. By virtue of this resolution the human essence becomes what it is, not by being an instantiation of an abstract universal, but by the creation of (new) social relations (VI Theses on Feuerbach).

Feuerbach's secularization of the self-alienated religious world was a clear attack on Hegel's canonization of the Christian cosmic drama. Like Kierkegaard's diatribe against Hegel in *Concluding Unscientific Postscripts*, Feuerbach was alert to the complications of understanding the human essence in a metaphysical sense. The projection of the human essence into an imaginary religious world, both philosophers argued, negates the constant striving typical of existing individual consciousness.[16]

The complacency imbued in the moment, in the Hegelian speculative drama, presupposes a completion that is impossible as long as "the subject is in existence."[17] More importantly, the finality that the subsisting achieves in the Hegelian system contradicts the knowledge that comes with existence: that the moments that constitute my life can be distinguished from one another and as separate do not constitute a system. Kierkegaard's challenge to the concept of an existential system, especially, reflects the difficulties with which the historian, as well as the philosopher, must contend when a "piece" of existence has moved into the past. The dilemma for both the historian and the philosopher is that the existing individual remains a part of the past which is the object of understanding. This requirement of existence, as Kierkegaard put it, negates the possibility of achieving a vantage point outside existence. To imagine the possibility of achieving a vantage point within and outside existence is only a deception perpetrated by the religious myth. Kierkegaard and Marx, it can be said, were both alarmed that without the demythologization of the medieval cosmic drama, human activity, at best, could only be conceived as the reflection of a systematic finality. The religious, medieval worldview resolved into its secular basis, however, acknowledges the omnipresence of the historical point of view which is synonymous with Ethics in Kierkegaard's words, and sensuous human activity in the Marxian critique. With the systematic finality of Hegel's philosophy, Kierkegaard was quick to point out, there is no room left for the existence of the "Herr Professor who writes the system" [18] It was this abrogation of individual existence and effort that led Marx and Kierkegaard to their mutual critique of Hegel's quietism.

The constant presence of the eternal in Hegel's scheme represents the systematic unity of Thought and Being, subject and object. Such a unity negates the interaction of the subject and the object. With no space between the subject and object, thought and being, there is no room for invention. With this form of tautological presence, the neat identity between subject and object, it is only thinking that is affirmed, not existence.

Intent on avoiding mere approximate knowledge, Hegel confirmed Reason as both the form and content of existence. There could then be no debate that what existed was the true reflection of rational consciousness. This peace that Hegel sought with reality was the crowning achievement of rational thought. This teaching, however, nullifies the immediacy of personal existence. The acquaintance we have with our own existence does not merely inform us of what exists, but equally sheds light on future possibilities. When my feet are cold, I am aware of a certain condition. If the condition persists, I am likely to take steps to overcome a discomfort. This personal knowledge is philosophy as the thought of existence, not the thought of the world, as Hegel would have it. It does not come too late to disclose possibilities, and it needn't wait for reality to mature before it can be understood as possibility. The skepticism that Hegel intended to avoid was intolerant of the notion that what subsists is not precisely the rational. If the result of this is that thinking becomes a form of deification, then the trade off is too great, since ethics is excluded and the imaginative interaction with nature and history is quashed. Marx and Kierkegaard found an appealing host in Hegel, for his systematic enterprise highlighted the drawbacks to the traditional conception of Thought and Being. Existence and consciousness lose their naturally inventive force and must be recalled, for the price of systematic finality is the loss of individual existence. Kierkegaard is more humorous in his efforts to dismantle the philosophic voice that speaks for all humanity, and he would ask Hegel to desist with the ploy that he could fly, even if he did jump higher than all previous philosophers. Marx, on the other hand, found no humor in the self-estrangement perpetrated by the religious myth.

The tautological condition of Hegel's system denies the historical point of view, since thought is restricted to a reflection of what is. Reality thus takes on a completion in the same way that the past is understood to be over. Thus conceived, reality is never subject to praxical modification, and tradition claims a natural ontological status. Hegel's world, as Pietro Piovanni puts it, "is a perfected cosmos, the monism of which has been amplified by the insertion within it of a scientific knowledge of new contents," which, of course, are always reduced to cognitive concepts.[19] What appears to be a changing world, then, can be reconstituted on a conceptual basis, and so Piovanni rightly concludes that Hegel's world is not the human world replete with plurality and non-reducible individualities.

Marx embraced Hegel's philosophy of becoming, but he shunned the primacy of logical development wherein every historical instance has a pre-ordained spot in the unravelling of universal history. He would agree with Vico that the logic of history is self-development illuminated by the power they both identified as invention. Vico recognized the reality of the self-making of history and, unlike Hegel, he stressed the origin of historical reality in contrast to a final goal of history. In this way he is aligned with Marx who also strove to delineate the origin of social relations in the sensuous activity of humanity.[20] Marx and Vico liberate history in their mutual humanistic emphasis on invention and free productivity, while Hegel conceptually entombed history in a transcendental logic. Marx and Vico, in their emphasis on the origin of historical phenomena, emancipated real history from sacred history. This inversion of the Hegelian system, as it is sometimes called, promises not merely a reflection of man, in Marx's words, but an understanding of man's place in the real world. This turn entails the realization that humanity, in fact, creates religion.[21] Hegelian transcendentalism denies the insight that humanity requires constant interaction with nature if it is to be realized. Moreover, Marx recognized that the character of human life meant that one's life is always an object, in the sense that it is a result of activity. The quietude of Hegel's worldview conceals the world of human activity and consequently real human consciousness is disavowed. The connection Marx drew between productive activity and the structure of the State, especially, was deeply antagonistic to the quietistic isomorphism of Thought and Being in Hegel's conception of history, which, he thought, disregarded real relations and limited itself to "high-sounding dramas of princes and states."[22]

Hegel, however, did not restrict his understanding of history to a theory of eminent persons and cosmic romance. His delineation of the approaches to writing history in some ways deviates from the more "high-sounding" drama of consciousness embodied in the *Phenomenology.* A discussion of his categories of historical methodology is reserved for a later discussion in which Gadamer, as a proponent of a subjective hermeneutic perspective, resuscitated features of Hegel's cosmic history. The compatibility of their seemingly antithetical interpretations of history will turn on the concept of tradition as the purveyor of the ontological condition.

The intersection between Marx's contribution to our understanding of historical reason and Martin Heidegger's interpretation of history lies in their shared view of human nature. Heidegger's challenge to conventional Western metaphysics, like Marx's, turns on his critique of the identity between Thought and Being, subject and object, that typifies Hegel's systematic philosophy. Heidegger's innovative ontology, put simply, begins with the peculiar ontical presence of human existence that simultaneously embodies an ontological dimension. The folly of Western philosophy, according to Heidegger, was the preoccupation with proving the existence of nature or external reality. The peculiar ontical character of human existence, he expressed in his early philosophy, comes with the immediate data that being implies being-in-the-world. Not in the spatial sense, but in the sense of dwelling, staying or sojourning. As a fundamental characteristic of human existence or *Dasein*, thinking always implies a world-context and, consequently, artifacts which engage consciousness. Within this nexus of being, human existence is distinguished in much the same way that Marx classifies it in *The Manifesto of the Communist Party*. Much as one might be horrified at the thought of abolishing private property, a vital stage in the culmination of the classless society—the undifferentiation of subject and object on the social plane—this stage embraces a filial attitude calling for the abolition of existence conceived as a commodity.

Marx's historical materialism does not discard the inner complexity of Hegel's dialectical transcendence, but it does offer us a real sense of the genuine productive forces at work in history. With the abolition *(Aufhebung)* of the feudal system, for example, the new order that was introduced "resolved personal worth into exchange value."[23] The labor theory of value (that is touted in Locke's treatise on government as the basis for acquisition of private property) is regarded as the source of "egotistical calculation," which encourages the "brutal exploitation" of labor and ever-expanding world markets. This attitude, which is identified as the basis for freedom in some economic circles, is the license to reduce all relations to money relations. This, in turn, resolves human existence into a commodity to be sold on the open market. Marx did not bemoan the abolition of such a social nexus, since such a culture encouraged people to function as machines.[24] Marx's critical look at the family is an example of the reduction of human existence to instruments of production; bourgeois marriage, in fact, is really prostitution on the private level. Heidegger would concur with Marx's conclusions on the evils of reducing humans to machines in his assessment technology's colonialization of human existence.

If there is a simple way into the complexity of *Being and Time,* it may be through the opposition Heidegger identified between inauthentic (*uneigentlich*) and authentic (*eigentlich*) existence. Much of Heidegger's challenge to traditional metaphysics rests on this elementary distinction. Heidegger, unlike Marx, is guarded about any prescriptive claims, since his primary interest in *Being and Time* is to outline a fundamental ontology, but we can not help drawing the conclusion that authentic existence is preferable, as it offers insight into the totality of *Dasein*, that kind of completion required for ontological grounding.

## Heidegger's Histo-Ontology

Heidegger's analysis of time-consciousness brings us full circle to Augustine's perspective on time. Heidegger speaks from the vantage point of a neo-Kantian. One of his earliest philosophical influences comes from Windelband and Rickert who outlined the difference between conceptualization in the historical sciences and the natural sciences. This distinction had previously been expressed in terms of the conceptualization of the real and the conceptualization of the general by Georg Simmel. His adaptation of these neo-Kantian influences, however, echoes the voice with which Augustine speaks of the movement of time in the *Confessions.*

Crucially important to Rickert's and Windelband's program was the understanding that conceptualization must accommodate the individuating historical mode, though they did not dispute the relevance of conceptual validity and the uniform regularities of the natural world. The neo-Kantian postulate that all conceptualization begins with a value-relation (*Wertbeziehung*) is at the heart of the historicist imagination and was duly imported into America in the 1930s by Carl Becker and Charles Beard. The value relation at the core of Rickert and Windelband's theorizing was emphatically not ethical but derivative of one's own historicity, that is, a given historical context necessitating a procedure of selection and identity with a larger cultural ideal.

Beyond his inheritance of the revival of Kantianism, Heidegger stakes the same philosophical claim that Vico and Marx proposed: that the fundamental historicity of human existence supersedes the rational discourse of traditional metaphysics. In its place, Heidegger implies, in his early philosophy, and clearly recommends in his later philosophy, that the invention characteristic of metaphor and poetry are the agents most helpful in disclosing truth. This links Heidegger to the primacy of poetry in humanistic movements. The preponderance of philological and etymological arguments and demonstrations in his writing is a hallmark of the humanistic preoccupation with the revelatory power of words. On the surface we can identify the connection between historicity and poetic thinking, as they both embrace the priority of the particular in disclosing truth. More profoundly, however, the structure of historicity (understood as the temporal horizon through which understanding takes place) and metaphor entail a commingling of seemingly disparate contexts; the result of which is a systematic unity of meaning.

Heidegger's challenge to traditional metaphysics begins in *Being and Time*, as does his personal inclination toward metaphorical language. It is perhaps Heidegger's use of the metaphor of the *Lichtung* around which his challenge to traditional Western metaphysics and its accompanying rational methods revolves. He introduces the picture of the *Lichtung* (a clearing in the forest pervaded by light) as that image which captures the tradition's mismanagement of the question of Being. The logical foundation upon which traditional metaphysics rests indicates that its real interest has been in asking the root question "why" matters are the way they are. Accordingly, the methods of Western metaphysics have been guided by the single rational principle that everything has a reason. Conceived in this manner, metaphysics traps beings in a theory of truth based on correspondence or *adequatio*. The metaphor of the *Lichtung* was intended to resuscitate the question of Being, as it refers to that clearing which makes things visible in the first place. The logical structure of rational discourse fixes things on the basis of sufficient reason and avoids the real issue that Being has no ground but is in fact the ground that simultaneously reveals and conceals[25] The problem with this solution, as Heidegger certainly realized, is that in overhauling the language of metaphysics to conform to the presencing made possible by the clearing of Being might very well put an end to philosophical discourse as the West knows it. If Heidegger did not intend to overhaul the discourse and practice of philosophy, it is at least clear that the pervasiveness of metaphor, especially in his major work, is philosophical expression that is descriptive of the landscape he traversed.

In the essay *Vom Wesen des Grundes*, Heidegger revokes the conventional starting point of transcendental philosophy, since, as he argues, transcendence is not introduced in the face of an object, but is what naturally happens with the fact of Being-there (*Dasei*n).[26] To refer to *Dasein* as transcendent is redundant, since it is always comported toward surpassing something, namely itself, in a movement toward totality. This self-transcendent movement is synonymous with the ecstatic unity of *Dasein*, as outlined in *Being and Time*; that is, *Dasein* has access to its own totality through the transcendent medium of time which requires no objectified field. Transcendental philosophy, then, becomes identical to the understanding of human existence as a self-transcending temporal unity.

It is in Heidegger's treatment of modern technology that we become keenly aware of the Western tradition's monotonous compulsion to secure and fix reality, rather than to let it emerge as with the ancient emphasis on *techne*. Conceived as a mode of unconcealedness, *techne*, since it brings forth that which is not yet before us, is not involved with a demanding manipulation and fixing of the environment which is a putting upon of nature's potential store of energy, but a gathering together of a completion which prescribes the way in which completion is to be realized.[27] The *Lichtung* becomes the root metaphor that calls for an approach to existence that is not constrained by the realm of logic and inferential methods. The *Lichtung*, then, functions as the scaffolding upon which the self-transcending dynamic of *Dasein* continually creates.

The uncovering and unconcealedness that the *Lichtung* invites is integral to the poetic structure of the world. The social relations that modern technology has spawned, however, places human existence into such ordered and designated structures that it conceals the kind of phenomenological self-revealing in which objects achieve their own presence. Human existence is threatened with this same state of affairs, since human essence is interpreted in light of the ordering function it possesses in relation to reality. Following Heidegger, the ordering compulsion of technology commits humanity to the status of "standing-reserve."[28] Tied to reality in this way, all other revelatory possibilities are abandoned. This is antithetical to that mode of revealing Heidegger identifies with an inventive causing-to-be or coming-to-pass which is also analogous to the creation of a poem (*poesis*). Presumably there is a fundamental affinity with poetic expression and the type of being possessed of human existence. We should expect that this mode of being is open to revealing itself, and so its ontical character will possess a fundamental ontological disposition. This is a consideration that is at the root of *Being and Time*; which will necessitate that human existence be distinguished from that kind of being that is ready for use as with the "standing reserve." Distinguished in this way, *Dasein* is always confronted with the possibility of creating itself.

There is a dual threat, then, to the compulsive demand technology has on reality: (1) alternative arrangements with nature are forsaken and (2) humanity runs the risk of becoming an extension of this compulsion and consequently a "standing-reserve," though humanity persists in thinking itself the ruler of nature.

The relationship between human historicity and a metaphoric philosophic voice is not peculiar to Heidegger. It is the philosophical voice that some scholars like Ernesto Grassi and Paul Oskar Kristeller associate with the Renaissance. The Renaissance is customarily associated with the revival of Greek and Roman Antique values, and thus it appears to have spoken with a borrowed voice. Moreover, the Renaissance's affiliation with the vague cultural movement known as "Humanism" has not helped to identify a unique philosophical voice. Still, the humanists' preoccupation with rhetoric, philology and the arts to the exclusion of metaphysics and logic have identified it as a primarily literary movement. Though there was little new philosophy to emerge with the Renaissance, a new-found individualism gave rise to a personal philosophical expression. The proliferation of letters incorporating the voice of the "I" meant that a new worldview had emerged from a period preoccupied with the mere preservation of knowledge. Petrarch's voice, for example, is identical with his communication; his recording of his personal thoughts and identity with Augustine on his trip to Mount Ventoux typifies this form of expression.

It was Dante more than any other person, according to Grassi, who demonstrated the relationship between the poetic word and the historical world. Grassi identifies Heidegger's metaphysical use of the *Lichtung* with Dante's proposal that language possesses an opening or clearing function. No doubt, Grassi concludes, Dante's fascination with the new dignity given to language coincides with the emergence of the vernacular and a revived appreciation for the many Italian dialects.[29] Grassi's claim that Dante believed the poetic word to be the way into the historicity of existence makes Dante a forerunner of Vico and Heidegger. We will return to the tantalizing connection between poetry and historicity in the final chapter, but now Heidegger's contribution to the critique of historical reason should be fleshed out.

The connection between the early and later philosophy, the appeal to a fundamental ontology and the assessment of language as "the House of Being," is centered on Heidegger's understanding of having an emerging totality-in-view, which in turn reveals the manner of appropriating possibilities. This issue is first addressed by Heidegger in *Being and Time,* chiefly elucidated in Division II, though the temporality of human existence is implied in Division I which is primarily occupied with analyzing the ontological constituents of human existence. As Heidegger proceeds with his enumeration of the categories of human existence, *Dasein* appears to be a many faceted ontological creature. The conclusion to Division I, however, that *Dasein* is fundamentally constituted of Care (*Sorge*) indicates that *Dasein* is essentially a unity. Division II is chiefly dedicated to explaining how human existence is possessed of a unity. The issue turns on Heidegger's distinction between authentic (*eigentlich*) and inauthentic (*uneigentlic*h) existence, a crucial distinction as the original German words suggest an owning and disowning of one's existence. After carefully distinguishing human existence from the modes of existence typical of things to be used as tools or things simply present at hand, *Dasein* can only be seen as something that is in each case my own. Heidegger's insistence that *Dasein* is in each case my own is reminiscent of Marx's admonition that the individual must not be conceived as an instantiation of an abstract universal, or more simply, as one of a type or "dumb genus." Heidegger's analysis of the constant temptation to fall prey to the inauthentic life, as he puts it, leads to an estrangement *(Entfremdung)* that is the foremost symptom of the alienating religious essence in Marx's writing. How, then, does Heidegger interpret the structure of Care (*Sorge*), and what relevance does it have to his understanding of temporality and time-consciousness?

Curiously, it is Heidegger's analysis of death that best illuminates Care as the unity of *Dasein*, authentic existence, the totality of *Dasein*, the future as the primary temporal mode of human existence, and finally what constitutes the genuine historical perspective. Since Heidegger's innovative analysis of death is the hub of the existential analytic of what distinguishes human existence from all other types of being and what distinguishes authentic existence from inauthentic existence, it serves as a way into his special understanding of history. Not history simply as what happens, culturally speaking, or what history is as an account of what has happened, but history as the happening of *Dasein*. This is the peculiar aspect of Heidegger's analysis of history that dovetails nicely with Augustine's meditation on time; that is, human existence is the source of historical movement, and it is not, as inaugurated by Augustine, a movement through space. Indeed, Heidegger insists that *Dasein* does not happen spatially, but in a manner more like travelling; his repeated metaphor that thinking is like moving along a way, perhaps the path that was cut by someone removing timber from the woods captures the movement of thinking.

The problem that Heidegger begins with in Division II of *Being and Time* is to clarify the primordial structure of that kind of being where Being is always an issue. This is a requirement in his anticipation of a fundamental ontology, and it necessitates having a grasp of the totality to be interpreted, since we must expect that any hermeneutic situation requires an apprehension of the totality of the entity being interpreted. In the case of *Dasein*, whose basic structure is that of Care, unity and totality are confused, since the Care structure means that it is always ahead of itself as potentiality-for-Being. The difficulty, then, in pursuing an interpretation of human existence is that it is so constituted that it is not-yet complete. Some form of closure is required to pursue any further analysis of human existence, and Heidegger proposed that end which closes off anything outstanding, death, as that limit which determines totality for *Dasein.* Death cannot be construed as a mere end point, since beyond such a limit *Dasein* is no longer possible, so an understanding of death must be outlined which suits the kind of being possessed by *Dasein.* Such a conception of death must correspond to the potentiality of *Dasein* as a totality, so it must take its departure from within the Care structure of *Dasein* which is always ahead-of-itself. Completion, *vis-à-vis Dasein,* is a matter of comportment and not the matter of a coherent summation that is realized after the fact.

As ontologically constitutive of *Dasein,* death is not a mere end point, but a being-towards and thus *Sein zum Tode* is that inner projection that will facilitate a view of the totality of *Dasein.* What Heidegger finally intended with this existential interpretation of death was that it is necessarily tied to the temporal structure of *Dasein* that is always-ahead-of-itself. Heidegger's interpretation of the fundamental structure of *Dasein,* as temporality, will not simply make intelligible the historical footing of *Dasein.* As the foundation for its Being, the historical will illuminate the natural inclination of *Dasei*n toward the exploration of history. [30]

Though an analysis of the inauthentic and the authentic modes of existence are inseparable, Heidegger's analysis of death points to the authentic as more revealing of the ontological character of *Dasein.* Why this is so is made clear in the common sense understanding that death signifies an ending that is in each case my own, not someone else's. That is, as a particular possibility, it is never a possibility that can be shared with someone else or somehow represented by someone else. As constitutive of *Dasein,* then, it signifies a possibility that is in each case my own. How dying as a possibility brings a totality into view is yet another matter, but *Sein zum Tode* reveals a possibility that ontologically provides a basis for authentic or *eigentlich* existence. Dying as a possibility, since it is attached to my existence at birth, means that *Dasein* already possesses its totality, if totality is what is subsumed within the beginning and end of an entity. In the case of *Dasein* the totality will encompass an intentional continuity derived from the projection of possibilities. This conclusion was only Heidegger's way of saying that human existence is forever comported toward its own completion, and so any understanding of human existence necessarily entails grasping totality or completion. Socrates's appeal to know thyself is complicated by this philosophical understanding of human existence, since it requires knowledge of a self-creating entity. How Heidegger begins to resolve this perplexity is through an analysis of that temporal continuum which is the ground of this process.

*Dasein,* then, possesses a continuity whereby what is not yet is a possibility that is really present, as with those entities to which becoming is constitutive of their character.[31] Heidegger's example of the fruit that possesses its ripeness, while still in its early growing stages, is typical of this kind of entity. *Dasein* is likewise construed, as it has access to its not-yet with the possibility of dying always before it. In this respect, Nothingness can be construed as the flip side of Being, but more importantly, the ownmost and insurmountable "not-yet" constitutive of *Dasein* is always something to which *Dasein* is comported. As projected towards an end, that which comes with the existence of *Dasein*, we would expect such a given possibility to inform *Dasein* in a special way. The impending possibility of death, in fact, is accompanied by an attitude that constantly challenges *Dasein*, and at the same time clarifies the inauthentic and authentic modes of existence.

If death, which is always essentially mine, is reduced to a public phenomenon, as Heidegger's analysis goes, it becomes perverted by the "they" which continually tempts *Dasein* to regard death as always happening to someone else. Basically, then, the everyday mode of existence, characterized by falling prey to the indefinite existence of the "they" attempts to evade the one possibility that is certain at birth. The real threat of this concealment is that being-toward-possibility and being comported toward an end is obscured. The routine evasion of death is psychologically institutionalized to overcome anxiety in the face of death, but at the same time *Dasein* has opted for a certain disposition toward death. Though the "they" advocates an inauthentic approach to death, one that constantly puts it off onto someone else, it also informs us of another option, the authentic mode which acknowledges it as a possibility. The authentic disposition toward death is not a morose brooding over an impending demise but a running ahead to a possibility.[32] There is nothing to be actualized in the event of death, for it signifies the possibility of impossibility, but the activity of running ahead allows *Dasein* to grasp that possibility that individuates. This liberation from the "they" is tantamount to discrediting the comprehension of human individuality as an instantiation of a universal abstraction. But, in being free for one's own death, a factical possibility reveals the ontological inclination of being-toward-possibility. The following inference is suggested in such an anticipatory outlook: not only is a particular factical possibility acknowledged, but the acknowledgement of this possibility, which signals a finality of sorts, sets up all the other possibilities that lie before the uttermost possibility of death.[33] Theoretically, then, a resoluteness towards one's ownmost possibility, in enclosing all other possibilities, incorporates the totality of *Dasein*. This grasp of a unitary structure meets the most important demand of any hermeneutic situation: that the totality of the entity to be interpreted be brought into view. This demand is guaranteed from within the basic structure of *Dasein*, and is further confirmed from a position of not having actualized an outstanding "not-yet." Though it is never to be actualized as a completion, since upon completion it is no more, the whole of *Dasein* can be presupposed in advance.[34] The hermeneutic situation, then, is (1) integral to the ontological constitution of human existence and (2) it is historical in character, since the totality of *Dasein* is a possibility that can be presupposed from the vantage point of "not-yet" being complete at some point in the future.

An awareness of time that is not an objectified thing but a temporalizing of temporality, or a generation of a time continuum, is revealed in the resolute anticipation with which *Dasein* faces its ownmost possibility. Being free for death, then, constitutes a model for authentic existence, and indicates that the future is the primary facet of time for *Dasein*. How a temporal totality is activated from anticipatory resoluteness has been implied, but the full scope of this temporalizing activity must be clarified, as it will explain why historicity becomes the original disposition for all forms of understanding. What does become clear, though, is the consequence of being free for one's own death as the model for authentic existence. In following what Heidegger calls the "call of conscience" and turning away from the variety of institutionalized ways of concealing one's ownmost possibility, *Dasein* becomes focussed on its limiting-situation[35] and attests to potentially being-a-whole by resolutely taking action.[36] The meaning of human existence, then, is not something that is confirmed from external sources but is evident in the self-understanding or self-projection of *Dasein* itself.[37] This self-understanding first comes to *Dasein* from the future to which it comes toward itself. How anything is to come toward itself presents immediate complications, but this problem is addressed in Heidegger's consideration of *Dasein* potentially constituting a whole at all times. This totality begins with a primordial future orientation, but the future as a facet of time never appears on its own, and is never something that is present as something at hand.

The structural whole or the Care structure of *Dasein* is first evidenced in the dynamic rapture that is initiated from being authentically disposed toward the future. This futural disposition simultaneously makes *Dasein* aware that it has been thrown into a situation where it faces decisions; the most clarifying element of its structure is the choice between authentic and inauthentic existence. The authentic future, then, takes over the condition of being tempted by the "they" and thus the resolute anticipation of *Dasein* returns to what has been. Consequently, the past or "having been" arises from the future. The state of being present is also disclosed in this dynamic rapture, for in taking action *Dasein* becomes circumspectively involved with one's environment. (*Umsich*t and *Umwelt)*[38] The totality that is *Dasein* is a temporal unity which in terms of the concept of Care is designated as Being-ahead, the future, Being-in-the-world or thrown, the past, and Being-alongside entities, the present.[39] Each facet of time is not to be regarded as a now or a space which can be filled, for they are mutually arising and as such they are not like entities, though they have a phenomenological status. They are located in the process of temporalizing and they are activated in their mutual arising. They are the preconditions for the most fundamental transcendent state in which *Dasein* is simultaneously outside-of-itself and for-itself, which makes the Care structure possible. These conclusions are especially remarkable, as *Dasein* is certified as essentially historical; not in the way that events occur in world-history, but in a more primordial fashion which makes historical understanding possible. This claim anticipates the fundamental difference between the ontical and the ontological in Heidegger's analysis of the historicity of human existence, since history is original to the constitution of being-in-the-world. Human existence cannot simply be regarded as a fact occurring in history, it provides the original temporal structure within which history as concern for the memory of past events and concern for the future becomes conceivable.

History as a science, on this analysis, can not amount to a mere summation of facts, if it is derived from the ontological assertion that *Dasein* is historical in its very nature. History, as a preoccupation of human existence, means that *Dasein* naturally occurs as circumspective concern within which it uses time. Time-reckoning, which is constitutive of being-in-the-world can only mean that all other phenomena encountered in the world will be accessible within time.[40]

The world is disclosed through the production of time which is at the seat of *Dasein*, but the temporal quality of things in the world, which is of an ontical character, really derives from this temporalization. Human existence and history are reciprocal, and therefore ontology and history are fused together at a pre-ontological level as proposed in Heidegger's use of the fable about Care's primeval shaping of the first human creature.[41] As the myth communicates, Care is breathed into the design of human existence in a posited mythological time and remains with it through its stay in this world. Heidegger's analysis of the care structure of *Dasein* is a philosophical elucidation of this ancient fable. Heidegger, in a footnote, acknowledges his debt to Goethe for his conceptualization of Care, which is found at the end of *Faust,* Part II. Goethe's character of Care, in turn, derives from antique literary sources, particularly from Virgil's *Aeneid.*

For our purposes, the result that human existence and history adjoin each other must be the most challenging conclusion of Heidegger's analysis. This reciprocity and the consequent historicity of *Dasein* can be viewed within the context of the historically minded philosophical school to which Heidegger owes his original impetus for the exposition of time. The next chapter explores in detail the consequence of the ecstatic temporal outlook intrinsic to *Dasein* within the historiographical framework which sponsored an analysis of time more applicable to the human studies. As heir to the historiographical tradition which challenged Ranke's more conventional view of history, Heidegger's proposal of time, as the transcendent horizon initiating a circumspective historical vision, challenges the tenability of historical reconstruction. The grounds for this controversy stem from a consideration of the past, that to which historical activity is directed, as an aggregate of ontical facts or events that are over and done with, that is, they possess a discrete completion. This characterization of history, that is, the past as transpired event that can be recalled through various forms of historical criticism, to which Ranke subscribed and promoted in his famous dictum *wie es eigentlich gewesen*, is undermined in the consideration of time as the dynamic temporalizing from within the self-understanding and self-creating of human existence. If self-understanding is the source of historical understanding, then the past will always be confirmed by an acknowledgement of the present, and if the present is really derived from the anticipation of the future, history must entail the ecstatic temporal unity that Heidegger identified as Care. If time-reckoning, even in the most ordinary ways originates from Being-in-the-world, then *Dasein* is always engaged with-in itself when engaged in any and all practical activity. As the model historical entity, then, *Dasein* is always directed toward making sense of the temporal ground from which it emerges.

# 2

# The Symmetry of History and Philosophy in German Historiography

## The Historiographical Origins of Effective-History

The rise of historical thinking in the nineteenth century gave birth to three major history journals in Germany, England, and France. Among the three, it was the German periodical that was most outspoken about its philosophical intentions. In the preface to the first issue of the *Historische Zeitschrift* (1859), Heinrich von Sybel announced that the journal's objective was neither antiquarian nor political, but would be committed to historical material which still had a vital link with present-day life. Sybel, who had disputed with his teacher Leopold von Ranke on political commitment in historical research, had decided that history, which recognizes the unity of all life and the "lawfulness" of life's processes, is the expression of the past which is contemporaneous with the present. This emphasis on the living facet of the past would be methodologically delineated in the neo-Kantian revival at the turn of the century. Moreover, what first emerged as a methodological concern would subsequently be elevated to a philosophical worldview.

Benedetto Croce articulated a similar historiographical position within the context of idealistic philosophy of history. His dictum, that "all history is contemporary history," equally expressed the force of the immediacy of the historical observer, and the consequent destruction of strict objectivity in the writing of history. Like the historiographical intentions of Sybel and the development of this theme in German historical circles, Croce's position redressed the Hegelian identity of history and metaphysics in which the hermeneutic impulse of the existential observer is squeezed out of the historical process. The position struck by Sybel and Croce has not been restricted to the confines of historical speculation. The obvious living force expressed in Sybel's account of history would pass through historians and philosophers alike. In both cases, each discipline has looked to the other to clarify their respective theoretical foundations. This is most clear in the special reciprocity of philosophy and history in Heidegger's particular ontological rendering of the historical theme which was first expressed by Sybel.

Heidegger's account of the real theme of history in *Being and Time* is predicated upon the reciprocity of doing philosophy and writing history. It is a curious position, since it too aligns history with metaphysics, in the way that Hegel demonstrated in his ontologizing of history. However, it does allow for the influence of the existential moment in both the writing of history and doing philosophy. In doing so, Heidegger has argued for the credibility of moving from empirically grounded discoveries, in historical methodology, to the more profound enterprise of ontological speculation. In what follows, I attempt to trace the representation of history as contemporary history in German philosophy and history from its fervor in the early *Historische Zeitschrift* to Heidegger's special ontological interpretation of history. The exposition of this special theme in history demonstrates the symmetry of the anatomy of historiography and doing philosophy.

Heidegger's view on history was not *sui generis*. He was deeply impressed by the neo-Kantian efforts to articulate the nature of methodology peculiar to the historical realm. This influence demanded the bifurcation of the historical and natural worlds. Heidegger's position, in this respect, went beyond Rickert's and Windelband's delineation of the nomothetic and idiographic methods, for it focused on the peculiar object of inquiry in the historical world– *Dasein* as a cultural-creating entity. Thus, he was more in line with Dilthey's aspiration of a critique of historical reason, though he would frown on Dilthey's despair at the cumbersome relativism of the historicist worldview. Rickert's position was partially inspired by the precocious work of Emile Lask, a student who was killed in World War I, and was presaged in Georg Simmel's *Die Probleme der Geschichtsphilosophie* (1892). Simmel provided two fundamental ideas that would be recast in Rickert's work,(1) the distinction between history as a science of reality and the natural sciences as the science of conceptualizations and generalizations, and (2) the neo-

Kantian liberation of the metaphysical subject in contrast to the metaphysical dominance of the objective world, a replication of the so-called Copernican Revolution in Kantianism.

In grounding history and philosophy in the special historical nature of human existence, Heidegger left unresolved the difficulty presented in leaping from the inherent evaluative structure of *Dasein* to ontology. He would insist that the ontological status of history, in the being of *Dasein,* did not entail prescriptivism, but the body of his work suggests a deep relationship between the value stance of the historian, first expressed in his *Zeitbegriff* essay of 1915, and the Care structure of *Dasein* in *Being and Time*. That there is some equivocation on this issue, in spite of his statements to the contrary, compels us, once more, to examine the validity of basing ontological speculation on the empirical enmeshment of the historical sciences. On the surface, the historiographical tradition that Heidegger's philosophy of history rests on demands that the subjectivism intrinsic to historical judgment be scrutinized within the explanatory framework of historicism, not naturalism. The criteria, then, that most fully explain and express the symmetry of history and doing philosophy embraces the existential premise that the combination of philosophy and history is not adequately represented within the naturalistic framework, since doing philosophy and history necessarily entails self-inquiry. The hermeneutic impulse of self-inquiry derives from the individual character of historical phenomena, which in turn enlivens interest and value as set forth by Rickert and Windelband. In contrast, the natural sciences aim at the general. Consequently, laws, not values, are generated. Gadamer's rendering of this hermeneutic position is expressed in his concepts of *Wirkungsgeschichte* (historical efficacy or effective-history) and *Horizontverschmelzung* (the fusion of horizons), and, like Heidegger's position, owes a sincere debt to Dilthey who previously expressed the difference between the historical and natural realms on the basis of different objects of inquiry. The hermeneutic consequence of Gadamer's position agrees with Heidegger's phenomenological interpretation of understanding, since historical development is formed from the inherited commitment to tradition and the corresponding grasp of future potentialities. The hermeneutic situation is really effective-history in operation, since understanding must always be historical.

The fusion of horizons in Gadamer's position recalls Schleiermacher's romantic hermeneutics in that historical understanding always involves "the placing of oneself within a process of tradition in which the past and present are constantly fused." (Dilthey also adapted Schleiermacher's procedure of *sich hineinversetzen* in his expression of *Nacherleben* and *Nachbildung*, with an emphasis not on mimesis but on establishing the continuity of the "I" and the "Thou"). Contrary to Schleiermacher's position, then, the real meaning of any objectification of mind, on the terms of effective-history, cannot be confined to the original source, since it is always partially determined by the situation of the interpreter. Consequently, the logic of historical understanding entails creation since it is really a productive attitude, and not merely restricted to reproduction. There are many reasons why the naturalistic model is deemed inadequate for the apprehension of the historical (civil/cultural) world, not least among them are the creative impulse inherent to interpretation and the individual character of historical phenomena. The individual character of historical phenomena, as first analysed by Simmel, was the basis for stimulating interest in the historical observer. Further, the creative synthesis inherent to historical inquiry of any kind entails the possibility of novelty and new formations, and so methods of inductive projection on the basis of observed regularities must diminish in significance as an explanatory framework embodying the continuity of the object of historical inquiry and historical thinker emerges. There is an explicit emergence that is analogous to the non-additive causality in evolutionary theory. This is the basis for the productive quality of historical reasoning which is embedded in the Kantian liberation of subjective consciousness. As Simmel enunciated, historical inquiry frees the mind from the empirical realm in the same way that Kant's epistemology unleashed the autonomous, transcendental ego in regard to the apprehension of the natural world.

Heidegger's account of the existential source of historiography and history in *Being and Time* demands the reciprocity of doing philosophy and history, since history's object of inquiry is continuous with (the "there" of) *Dasein* whose unique existence is characterized by Care, or the circumspective vision Heidegger signified with the term ek-stasis. This term is a play on the word existence which Heidegger used in place of the conventional concept of consciousness. The particular presencing of human existence, for Heidegger, was not adequately rendered in the term consciousness, since it did not convey the temporal movement integral to *Dasein,* nor had consciousness been interpreted to include a world and others. The regulatory nature of the Kantian subjective consciousness would devolve into concern for a world, and even others, in Heidegger's philosophy.

The continuity reflected in this special kind of ek-sistence is best expressed in the temporal posture of *Dasein*, but it had received careful attention before Heidegger's articulation of historical presence and historicity. Although what emerges as an ontological theme in Heidegger was addressed by philosophically minded historians, it began as a methodological concern in the historical sciences. This development offers much to the contemporary philosopher seeking the mutual influence of history and doing philosophy. It needn't result in a new philosophical genre that somehow blends the compatibility of two separate genres. Nor should it mean selecting one philosophical genre over another. If the mixing of history and philosophy presents a crisis in philosophical circles, it is only because the inevitability of developmental historical contexts, as the source of all thinking, and the historical unity of the subject-object relationship, as the ground for interpretation, is regarded as foundationless, in the traditional sense. Nonetheless, the narrative structure of historical accounts is not incompatible with philosophy that begins with an inclusive rendering of consciousness which in turn entails the circumspective and recollective facets of human existence. This is apparent in Heidegger's concept of *Wiederholung*, as we will see.

The neo-Kantian innovations in historical methodolgy were preoccupied with the pitfalls of historicism Simmel did not participate in the debate that concerned Rickert, Meinecke, and Troeltsch, since he died in 1918. What would emerge as a distinct solution was the establishment of the validity of a horizonal and shifting foundation seated in the consciousness of time. This naturally means coming to terms with the explicit historicism in all of philosophy. This needn't result in casting the net of relativism over all philosophy, though it might indicate the impossibility of an ego-less stance in philosophic inquiry. However, it must mean that historicism, conceived as a relationism, is a necessary component of all philosophical speculation. But this is nothing new to philosophy which has never occurred in a vacuum. The heritage of this philosophical insight can be recognized in the *physis/nomos* controversy associated with the Greek Sophists. This debate represents the original transition from concern with the uniform regularities of nature to an introspective analysis of the human condition that is always marked by specific contexts, intentionality, or a positive enactment or law as the word *nomos* indicates in the Greek. The shift from speculation of the natural order to debate over law and convention can be interpreted as the prototypical instance of distinguishing the natural scientific enterprise of discerning general laws from concern for human institutions and legislation which is clearly a product of choice and a social and cultural milieu.

## Heidegger's Predecessors: The Concept of Historical Causality in Meinecke, Troeltsch, and Others

Sybel's pronouncement in the preface to the first edition of the *historische Zeitschrift* implied that the lively and determining influence of the past upon the historical observer must be the starting point of historical investigation. If history begins with the past "speaking" to the historical thinker, then the link between past and present is mediated by the interest provoked by the effective presence of the past. Neither the *Revue Historique* nor *The English Historical Review* expressed such fervor over such a philosophical premise. Friedrich Meinecke, who later assumed the editorship of the *Historische Zeitschrift*, was equally insistent on the relationship of value to historical interpretation. Through his articulation of the causal forces in history, Meinecke arrived at the position that the essence of history involves not just causalities but value. Among the historical forces he identified: the mechanistic, the biological, and the moral-spiritual, the latter, only, could explain the purposive drive of personality and the spontaneous activity of people.

The nature of history, as with Sybel, constituted an inner continuity that deep down required no mediation of explanatory mechanisms. The link provided by the value-stance of the historian and the effectiveness of the past meant that "historical seeing" was truly an unmediated experience. As such, it would make sense in German to refer to such an experience as *Erlebnis,* as opposed to *Erfahrung*. This distinction, as promulgated in the *Lebensphilosophie* of Dilthey, has special meaning within the historical realm, where understanding is really the rediscovery of the "I" in the "Thou." Dilthey, in addressing the process of understanding in history, also intended the efficacy of historical movement that is mediated by the duration of the past into the present. More than this, though, *Erlebnis,* as the basic datum of history, was regarded by Dilthey as the given raw material for any kind of imaginative formation. The concept does not refer to an experience that must be inferred or confirmed by a secondary source, it is immediate and reflects a "seeing as" experience that is sanctioned by intuitive links with the past that momentarily completes an historical structure. When something is apprehended, in this manner, it is still alive in the way that the past persists into the present in Gadamer's concept of *Wirkungsgeschichte.* It was poetry, Dilthey concluded, that represents this kind of internal experiencing. He also singled out Goethe as the epitome of this form of experiencing life.

Heidegger's later philosophy, emphasizing the ontological structure of poetic speech, revived Dilthey's intuition, for it theorized that poetry grasped the primordial in such a way that it was not merely referential but stood for itself. This was a position held by Novalis and inspired by Hölderlin. Heidegger's later philosophy promised that poetry's independence from external determination would allow Being to speak, thus overcoming conventional metaphysic's preoccupation with talk of mere objective beings. The speech of poetry was not mere talk for Heidegger, it was disclosure and unconcealedness, the activity of truth that had been abandoned since the early Greek's preoccupation with substance as living matter. Furthermore, the process of selection on the part of the historian, within the context of historical efficacy, naturally excludes some information while highlighting other occurrences. According to Ernst Troeltsch, this procedure, guided by the historically conditioned situation of the observer, and possibly directed toward an anticipated effect, for example, understanding the present or even generating action, encircles the historical observer in a nexus of consequences and effects (*Wirkungszusammenhängenden).* This nexus completes the view of the historical observer in the same manner that Dilthey's acquired psychic nexus (*erworbener zusammenhang des Seelenslebens*) informs the poetic imagination in its effort to enliven a certain image. Both processes complete a lived experience without necessarily absolutizing it.

Dilthey's and Troeltsch's conception of the creative posture of the historical observer enunciates the correspondence between the poetic imagination and doing history. This connection does not suggest the stylistic compatibility of poetry and history, but, more importantly, the more profound understanding that history and poetry are generated from the same imaginative faculty. In this connection of historical activity (efficacy), not only is new synthesis possible, but something totally new which did not previously exist can emerge. Spontaneous creation, originating in the "own connection of life" (*eigenen Lebenszusammenhang*) is simultaneously criticism and continuation. Troeltsch advocated a form of historical reason, effectively countering Hegel's *List der Vernunft*, when in this nexus of criticism and creation, the "final secret" is faith in a spontaneous reason *(momentane Vernunft*) and the will to affirm it. This "leap" is sympathetic to Kierkegaard's polemic against Hegel's "existential system," and, like Kierkegaard, he disputed the tenability of the Absolute from the historical point of view. Completion, though, remained a real possibility within the nexus that culminates in the contemporary observer's teleological continuation of the vital past.

Because of the deep imprint of the spiritual-ethical strata, according to Meinecke, pure naturalistic science could not adequately penetrate the historical realm. If history were comprised of a strict causal nexus, then the totality of human events would be its proper theme. But, since this has never been feasible from the historical point of view, unless one chooses to impose a strict logical formula onto existence, as Hegel did, then history necessarily entails a process of selection. Such reasoning led Meinecke to postulate another form of causality besides pure causal determination. Historical influence, as he put it, is more related to the enduring and profitable effects an original event precipitated. This supra-causal (*übercausale*) meaning of history, as it penetrates the practical realm of our existence, concerns life-values. "History gives us the content, wisdom, and signposts of our lives," he declared. In this respect, Meinecke expressed the compatibility of artistic creation and historical inquiry, since both pursuits are predicated on intuitive relationships. The historian, like the artist who seeks a more direct and living relationship with his craft, is directed by personal intuitions that reflect a direct relationship with historical phenomena. This was a theme that was expressed as early as Humboldt's thoughts on the subject in 1821, and it remained a vital tenet of historiographical speculation in Germany.

Meinecke, like Troeltsch and Heidegger, invoked Eduard Meyer's analysis of history in formulating his notion of effective-history and its creative persuasion on the present. Meyer promulgated the value of history as a living internal process emanating from the awakening of interest and concern in the contemporary viewer. When Meyer referred to history as "any event of the past the effect of which is not exhausted in its moment of appearance but clearly continues to have an effect upon succeeding periods and to generate new processes in them," he reiterated the sentiment that guided Sybel in the beginning days of the *Historische Zeitschrift*. In keeping the tradition alive that defined history and historiography as an interpretive enterprise, and thus a profoundly creative activity, he influenced the generation of thinkers, like Heidegger and Gadamer, who would complete this historical formula by ontologizing historicity itself. History, as contemporary history, is the original format for Gadamer's hermeneutics and the analytic of *Dasein*.

Paul Tillich, who reviewed Troeltsch's *Der Historismus und seine Probleme* in 1924, sensed that Troeltsch saw the need for a historical category between the general and the particular. It would be Troeltsch's category of individual totality that embraced both the individual and the general in historical interpretation. He had first introduced this category in his "Historiography" entry in Hastings' *Encyclopedia for Religion and Ethics* (1912). Troeltsch was not remiss in crediting Simmel with an earlier articulation of the a priori historical forms of selection, individuality, totality and meaning with which the neo-Kantians were concerned. He was later to expand on the central category of individuality in his formal logic of history in *Der Historismus und seine Probleme* (1922). The "Historiography" article also reveals his close ties with Meinecke's thinking. Both thinkers were certain that the historical world called for an explanatory framework other than that in the positivistic world of the natural sciences. Troeltsch matched Meinecke's claim for the effects of supra-causality in history with his notion of non-equivalent causality. Metaphysically speaking, this form of causality expressed the fundamental creativity of historical reality. For Troeltsch, no account of antecedent reality, within the framework of the static causal connection of positivism, could properly explain the emergence of the new and novel. Troeltsch and Meinecke, in analyzing the nature of historical causality, were equally concerned with the enduring thread of values that can be continually resuscitated from the past and transcended through the determination of future synthesis.

Tillich, who expressed his own anti-Hegelian position in *The Interpretation of History* (1936), in singling out Troeltsch's innovation of individual totality, had pointed to the deep constitutive presence that history has on consciousness. As Tillich remarked, Troeltsch was not merely invested in promoting a particular sense of history, he intended to articulate what it meant to stand in history. With Troeltsch one can readily see the influence of the categorial significance of history that would mark Heidegger's work.

Tillich found weaknesses in Troeltsch's category of individual totality, since Troeltsch also defined historical time as a living fusion, like a stream where nothing stands isolated. This fusion of all times and interconnection of all events made historical totality open on all sides, so to speak. Troeltsch, in claiming the power of individual contexts and the "own situation" of the historical thinker in forming historical totality, also opened himself to severe criticisms of radical subjectivism and mystical divination (as the interpretative tools of historical inquiry). Tillich reduced Troeltsch's category to the notion of *Gestalten,* and placed little emphasis on the process underlying Troeltsch's formulation of individual totality and historical individuality. Nonetheless, Tillich's assessment of Troeltsch's effort to resolve historicism indicates the undeniable effects of history on doing philosophy. To regard thinking as seated in history, as Troeltsch did, is to entertain the belief that philosophy mutually arises with history. Consciousness, in other words, is only conceivable within history. Heidegger's innovation of *Dasein* resolves consciousness into historical projection and is the philosophical analogue for Troeltsch's interpretation of the fundamental historical posture of human reality.

The way into history, according to Troeltsch, was through an apprehension and appreciation of the notion of individual totality (*individuelle Totalitätsbegriffe*). Any apprehension of the total course of events was cosmological romance, as Croce remarked about Hegel. The logically conclusive point of view, regarding the apprehension of historical wholes, was that they were not to be apprehended through a systematic addition of elementary occurrences, but through their intuitive unity and effective fusion with a "greater or smaller, historically significant totality of life." Here Troeltsch argued for, along with Meinecke, the intuitive sense of one's own personal reality in grasping historical phenomena. The indissolubility of the individual, in the apprehension of historical reality, was analogous to the intuitive and imaginative flare with which the artist guides the creative process.

By the time that Troeltsch compiled *Der Historismus und seine Probleme* he had successfully broken from the Platonic framework that had informed his philosophy of religion. He had also progressively distanced himself from Hegelianism. His predisposition towards historicism can be traced throughout his body of work. At the writing of his essay on William James' philosophy of religion one can discern, for example, his tenuous hold on the "religious *a priori*." The full expression of his historicism in *Der Historismus* undeniably reflects a reordering of transcendental philosophy. Where James had become radically empirical, Troeltsch had become radically historical.

Troeltsch's assessment of historical inquiry and the nature of the historical, wherein the incorporation of all individuals in a super-individual connection is fed by tradition and the past, led to his belief in the pre-eminence of the individual and the "moment of originality" of the historical spectator. From the depths of this personal originality the individual possesses the power to transform the whole (of history). Troeltsch asserted that this was no mere assumption but could be empirically verified through production. Metaphysically speaking, this logic– characterized as the "irrational logic of the new and creative"– was the underlying structure of all reality. It was the driving impulse of all transformation and assimilation, and represents the most important theme and tool for the historian.

The will to design, on the part of the historian, arising from the creative impulse of the enduring past reawakening a spiritual-ethical deposit of reality, was the decisive role of meaning and value in historical interpretation. The activation of this historical unconscious, Troeltsch argued, cannot be rationalized for it belongs to the circle of the momentaneous and original own situation of the historical thinker.

At this point in the discussion the correspondence between Meinecke's and Troeltsch's rendering of the special forms of causality in the historical world and the causality of freedom in Kant's philosophy becomes evident. Where Kant had articulated the division between natural causality and the causality of freedom in morality, German historiography had also elucidated the independence of the historical world. If Dilthey envisioned a critique of historical reason, we can expect that it would more closely resemble Kant's critique of practical reason which in its own way established the autonomy of creative human activity.

## Historical Connectedness, Teleology, and the Virtue of the Hermeneutic Circle

Troeltsch's logic of history, as with Kierkegaard before him, was inspired by a profound rejection of Hegelianism. The emphasis on the impossibility of apprehending historical wholes on the basis of a uniform law, naturalistic or otherwise, led him to articulate the primordial characteriological determination underlying the historical. Moreover, Troeltsch's analysis of the historical unconscious prefigured Gadamer's hermeneutics and, as with Heidegger, the interpretative circle surrounding the own situation of the historical thinker is deeply positive, not a vicious circle. That the creative stance of the historical thinker reflects more than methodology becomes most apparent in Heidegger's ontological rendering of historicism.

In this respect, Troeltsch went beyond Meinecke's interpretation of the contemporaneity of history. The characteriological determination of history, expressed in the will to design on the part of the historian, equally meant that history was only intelligible as the medium for the self-production of personality.

The most significant logical consequence of historical continuity– causally inexplicable and only comprehensible by a feeling into the unity of becoming– was that historical time could only refer to an inner space and sense which orients the past and present in service of the future. This sentiment, of course, was fervently argued for by Nietzsche in *The Use and Abuse of History*, who also endorsed Goethe as the epitome of historicist thinking. Not surprisingly, Nietzsche's insights into the facets of historical consciousness find their way into Heidegger's analytic of *Dasein*, where, in fact, they inform the historical vision of human reality.

As nothing is isolated, given Troeltsch's concept of historical time, the past and future are merged in a productive manner. Troeltsch's articulation of historical time was a continued effort to clarify the essence of history as something that just does not happen once. The living organic fusion of historical time (*Verschmolzenheit*) is not just an uninterrupted flow, but through memory and value the past and present can be "creatively united at every moment." Troeltsch identified this intuitive process of uniting history, from the vantage point of one's own situation, "historical seeing" (*das historische Sehen*). Not only did this "historical seeing" represent the real essence of history for Troeltsch, it was equally the real essence of thinking. To see in this way means that the experience of the historical is immediately given, so it cannot be that the simple past is apprehended. It must be that what is yielded by the past is not something that is merely left over, but something that is still alive, as illustrated in the concept of *Erlebnis*.

If continuous historical development can only be construed from the standpoint of the present observer–what we must above all accept in light of the contemporaneity of history– then the process, as articulated by Troeltsch, appears to be leading from the past to the present in an inner continuity. Troeltsch warns, in this light, that the teleology evident in this whole process is not to be understood in the ordinary sense, that is, with the spectator as the end of the whole. Rather, the inner connection, effective up to the present, stands before the observer as the past and present at the same time, that is, the present is illuminated from the past and the past is reflected from the present. This apparent antinomy is a necessary step in the elucidation of the unifying life-process (*ein einheitlicher Lebensprozess*) underlying historical inquiry and history. (Both Troeltsch and Meinecke would refer to the reciprocity of the whole and the particular in history as the *schaffend*er *Spiegel* [creative mirror] of historicism.) The certainty of being a result of the past and simultaneously the initiator of collecting and continuing the past is reconciled and eventually proved through the winning of a "conception of a material presence," which is a "unity of the heterogeneous." This result represents Troeltsch's major critique of Hegel's philosophy of history. What is gained in the reconciliation of the above antinomy is a concept of our "own situation"– a material philosophy of history, which is entailed by the logic of history. Troeltsch never fully articulated a material philosophy of history. His logic of history implies an activism in so far as the productive function of the historical observer is reflected in the assimilation and unification of the heterogeneous, that is, the given material of history. However, his logic does not represent a fundamental premise from which historical occurrences can be derived.

The implication that a material outcome is entailed by a logic of history pragmatically verifies Troeltsch's insight into the creative core of historicism. And, at the same time, it decisively acknowledges the autonomy of the historical realm in the manner that Kant guaranteed the independence of the moral realm. It was Troeltsch's intention, as well, to clarify the ethical character of historical creation (interpretation). The comprehension of the present as the result of the past total process is equally directed toward transformation, "in which not only the past and present, but the future is assumed." Troeltsch saw the tendency toward the general in the affirmation of the "own situation" as the culmination of a total unifying process directed toward future-formation. "As the spectator (historical observer) feels it out of his own situation," the development of the store of historical reality is directed toward that which should be. The inescapable conclusion of Troeltsch's logic of history is that development is "determined" by a *must* and an inner impulse to the general. Taking stock of the present, as "history as contemporaneity" dictates, never signifies a completed historical process, for history is also an account of the future to be determined through self-formation and self-criticism. This concrete-individual determination is simultaneously a *task* and a *must*, and: "With this must, pure descriptive and investigative history ends, and active history, or historical life begins."

Troeltsch paid his respects to the master Ranke in *Der Historismus und seine Probleme,* but ultimately agreed with Ranke's most severe critics, Sybel among them. Troeltsch sensed the value of the historical unconsciousness and felt deeply that history was never merely the reconstruction of past events. Ranke had been aware of the didactic school in historiography, but he modestly claimed that his work sought to uncover how things actually happened (*wie es eigentlich gewesen*). Many condemned Ranke's position as hopelessly apolitical, but, more importantly, his emphasis on the past and objectivity (*Unparteilichkeit)* eclipsed the vital role of the present position and interest of the historical thinker. In restricting himself to the reconstruction of past events, Ranke did not openly promote history as the living presence of the past in the present.

The historical unconscious became apparent in the historiographical themes of nineteenth-century philosophy. Together with the concept of time as duration, it distinguishes history as the fundamental structure of human reality. For Troeltsch, especially, it meant the existential union of value and method. The wedding of value and method in historiography was by no means restricted to German scholarship, though it invariably focused on Hegel's interpretation of history, as in the case of Benedetto Croce.

## Heidegger's Earliest Rumination on the Nature of History: *Der Zeitbegriff in der Geschictswissenschaften*

There can be, then, no clear demarcation between philosophy and history, and nowhere is this more apparent than in Heidegger's existential analytic of *Dasein* or human reality. This analysis disclosed the profound consequences of distinguishing between the human studies and the natural sciences. He articulated the difference in the manner that Troeltsch, Meinecke, and Croce revealed the status of time in historical consciousness. Although Heidegger favored the neo-Kantian framework at the writing of *Der Zeitbegriff in der Geschichtswissen-schaften* in 1915, the article pointed to the special unity between life and past deed in historiography. Like Meinecke and Troeltsch, he was influenced by Eduard Meyer's expression of the contemporaneity of history and the neo-Kantian innovation of the metaphysics of the subject, that is, the autonomous, productive and regulatory temporal self. Thucydides's contention that history should relate contemporary events originally expressed the ideal of an unmediated intake of historical causality. The effects of *das historisch Wirksame* is the result of the influence that a historical happening has upon the present. The past, then, only has significance when seen from the present.*(Vergangenheit hat immer nur Sinn von einer Gegenwart aus gesehen*), thus highlighting the pivotal role of the productive self.

Like Troeltsch's historical speculation, Heidegger's research has its roots in an understanding of historical methodology. In this early essay, one can discern the germ of the categorial significance of history, underscored in the final paragraphs of *Being and Time*. The correlation between the writing of history and what it means to be seated in history–exemplified in the existence of *Dasein*– indicates that transcendental philosophy corresponds to empirical discoveries, that is, in history and its methods. Heidegger, of course, accomplished this feat in his exposition of *Dasein* as that ontic entity whose special status is to be ontological. Here the tension between the historical view and the philosophical aspiration of foundations or first principles rests on the complex nature of human reality. Clearly though, philosophy comes to history from within the activity of history, it is not superimposed over history.

The original meaning of time in history, then, stems from the awakening of historical meaning when the qualitative otherness of past events presses into the consciousness of the present. This historically effective process would be later canonized as *Wirkungsgeschichte* in Gadamer's hermeneutics. Heidegger ended his *Zeitbegriff* essay by addressing the bifurcation of the natural and historical world. This was perhaps first expressed in historiographical circles in Gustav Droysen's *Grundriss der Historik* (1858) which originated from a series of lectures at the University of Berlin. This outline for history and historical methods begins with the delineation between the world of nature and the historical world, and then distinguishes history as a restless succession of continuity advancing upon itself. This characteristic of endless succession, or continuity, is the crux of the distinction between nature and history, since nature is comprised of endless repetition with no real new forms resulting from this process (evolution [which is really history] notwithstanding). History, on the other hand, as defined by Droysen, always involves penetrating a process of growth or advance that is always contributing to itself. (See chapter 5 for an extended discussion of Droysen's interpretation of history). This expression of the meaning of history is the core of the logic of history that emanates from the historiographical innovations in the nineteenth century. It resembles Hegel's logic in so far as this process of self-contribution means that ideally all past events are reflected in the new, but it differs from Hegel's history in that time, as the discrete form of history, swells up from within the material of historical continuity. As such, the form of history is integral to development, it is not the teleological category that subsumes continuous historical movement under an *a priori* ideal form. The adaptation of Kant's categories to innovations in historical methodology guaranteed only the regulatory, and thus creative, aspect of the *a priori* form of history.

Heidegger, in making the distinction between nature and history, cites a study on Augustine by Troeltsch *(Die christliche Antike und das Mittelalter in Anschluss an die Schrift 'de civitate Dei'*, [Berlin, 1915] 430–31). Troeltsch's study explicated the function of periodization in historical inquiry, thus recognizing the essential qualitative component in history. The main principle of historical concept formation is then revealed as value-relatedness (*Wertbeziehung),* according to Heidegger. Heidegger acknowledged his debt to Rickert's formulation of this issue in *Die Grenzen der naturwissenschaftlichen Begriffsbildung,* and therefore aligned himself with the neo-Kantian critique of Hegel's historical monism, which denied the living, existential value in the assessment of historical movement. Above all, the revelation that time computation in history is totally other than calculating time in the natural sciences established, for Heidegger, that historical inquiry is an original and autonomous attitude of the mind, irreducible to any other science.

Droysen, in his influential *Grundriss*, had stressed that time as the discrete form of historical organization should be methodologically segregated from the spatial categories of the natural sciences, which emphasise the existence of discrete and isolated elements. History, and the investigation of development, requires the simultaneous evaluation of the part and the whole, a hermeneutical posture that places historical investigation on other than positivistic foundations.

The claims of autonomy for historical investigation separates it from positivism and naturalistic interpretation. It further distinguishes history as the transcendental medium through which the past is tied to the present. Naturally, this must mean that the past is still happening. The process of effective-history might be an historical discovery by those concerned with methodology, but if it ascertains that time is the medium which encloses consciousness, its effects are certainly not limited to history. In declaring the independence of the historical attitude, which we come to know through the activity of effective-history, Heidegger also took the necessary first step in elucidating a critique of historical reason.

Friedrich Meinecke, thirty years before Heidegger's *Zeitbegriff* essay, had written his state examination on the comparison between the methods of the natural sciences and the methods appropriate to the human studies. In this context Meinecke was especially concerned with the question of personality. This concern was inspired by a Eureka experience during a lecture by Droysen. As recounted by Carl Hinrichs, in the introduction to the English translation of *Die Entstehung der Historismus*, Droysen, in a lecture on the Sistine Chapel, had argued that no mere addition of the natural and material elements involved in the painting of the Chapel could explain the final result without the addition of an "x" factor. This "x" factor was to be understood as the unpredictable element introduced by the willing, creative subject. This formative experience made the interest in personality and the creative individual a passion for Meinecke, as it was for Ernst Troeltsch.

Heidegger's early essay reflects the influence from Rickert and Windelband's innovations in methodology. But beyond this it demonstrates a more profound basis for the distinction between the natural and human studies. Heidegger implied that the two realms were to be distinguished on the grounds of the respective object of inquiry. This sensitivity to the object of inquiry in the sciences undermines any attempt to unify the sciences on the common grounds of methodology, as the object of inquiry will always demand a different point of view.

That Heidegger made use of Troeltsch's research is significant, but that Heidegger and Troeltsch looked to Heinrich Rickert for the delineation of the historical world is far more important. This influence reflects the more encompassing intellectual movement, the call back to Kant, that combined with methodological innovations in historiography, resuscitated Dilthey's aspiration for a critique of historical reason. In effect, emulating Dilthey's program intended to modify Kant's transcendental imagination into a historical unity of apperception. It should come as no surprise that such intentions would confirm Hegel's approach to history as "cosmological romance," and at the same time, elevate the virtues of the hermeneutic circle to methodological canon. This elevation entails the mutual reciprocity of the individual and the whole in history, and does not deny, as Hegel's approach to history does, the actual reciprocity of the particular and the general, the part and the whole in historical movement. There is a relationship between the intention of the Spirit and historical actuality in Hegel's history, but it takes the form of Reason imposed upon actuality. The result is the canonization of the subsisting by virtue of the powers of the *List der Vernunft*, it is not a mutual relationship in which historical actuality achieves either teleological completion or a rational status after the historical past is assimilated from the perspective of the "own situation" that informs the discretion of any given contemporary situation.

As we have already seen, such a positive link between the part and the whole in history was established in Troeltsch's category of individual totality. Heidegger also revealed the virtue of the hermeneutic circle in the same manner, since the totality of *Dasein* is realizable even when there remains something outstanding (most notably in relation to death). The ability to grasp the whole, when it is still outstanding, makes future possession a necessary condition of all experience. The possibility of conceiving the whole, even when there remains something outstanding, must be the primary discovery of the historical investigations of Heidegger and Troeltsch before him.

## Time, the Imagination, and the Care Structure of *Dasein*

Heidegger's *Kantbuch* of 1929 reveals his deep attraction to the Kantian critique of experience and consciousness. The book continues the reversal of traditional metaphysics inaugurated in *Being and Time* and sets forth Heidegger's most detailed analysis of the imagination. The domination of the subject by the object, perpetrated by traditional metaphysics, was transformed by Heidegger into a conception of Being grounded in the "possibility of experience." The quest for Being, as initiated by Kant, instigated a concerted move from a reality in itself to the subjective foundations of consciousness. Heidegger intensified Kant's regression to time as the fundamental intuition of all experience, since he delineated the existential details and consequences of such an intuition. Kant had promoted time as that intuition that governs consciousness of the self, as opposed to the apprehension of external phenomena. This platform was assumed by Droysen and others as the initial premise of historical investigation, since time was the essential ingredient in the discovery and penetration of the kind of advance that produced ascending transitions that continually are supplemented from within, Droysen's principle of "contributing to itself" (#2 *Grundriss).*

In the *Kantbuch* Heidegger reiterated two themes that had been previously addressed in *Being and Time*: (1) that time and the "I think" (Kant's *Ich denke* or the unity of apperception at the seat of consciousness) are not only compatible but identical, and (2) that the transcendental imagination, the primordial unity of the faculties of the understanding, is rooted in time. These themes, addressing Heidegger's most ambitious project, that is, the resuscitation of the search for Being through that entity most concerned with its own being, that is, *Dasein,* are noteworthy, since Heidegger had identified the real theme of history, not as something universal hovering over the world, nor the unique which only happens once, but as "possibilities which have been factically existent." This assessment of history amends both the Hegelian and neo-Kantian views of history, and introduces a new conception of history which can only be understood by an analysis of that being which "happens historically," that is, does not merely occur in history. This historical happening of *Dasein* is an existential example of the self-referential and self-contributing facets of history outlined in Droysen's *Grundriss.* Hermeneutically expressed, inquiry into history devolves into self-inquiry. Furthermore, Heidegger, unlike Kant, was absorbed by the distinction between the natural sciences and the human studies. Kant, on the other hand, uniformly applied a universal conceptualization of reason to his thinking, that is, the method of reasoning remained the same throughout his Critiques. The primary consequence of a grounding in the methodological distinction between the natural sciences and the human studies guarantees that temporality is exclusively the fundamental category of the subjective self. Furthermore, it established that we do not come into history, but that history derives from us.

Heidegger's investigation revealed that the historical status of *Dasein* makes it more than world-historical, in the way that other entities occur in history. As such, philosophy as a human enterprise, can only be performed historically. To grasp Heidegger's concept of history, then, one must first understand how Heidegger structured that existence most privileged to the court of Being, *Dasein*. That *Dasein* is constitutionally disposed toward Being explains Heidegger's metaphorical appraisal of *Dasein* as *die Lichtung,* the lit forest arena that invites a disclosure of Being, but we must look to the critique at the heart of historicized consciousness to initially assess this primordial openness.

By the end of Division I of *Being and Time*, Heidegger had given every indication that *Dasein* is oriented through historical projection. That *Dase*in is structurally constituted of Care, the conclusion to Division I, means, among other things, that human reality is always ahead of itself. This Care structure signifies, then, that a unity already constitutes the whole of human existence. This temporal structure is dramatically outlined in *Dasein*'s anticipatory resoluteness toward its own end (being-toward-death), and clarifies that *Dasein* is essentially futural. In freely facing death, a possibility that has been inherited (one is old enough to die as soon as one is born), *Dasein* is thrown back upon its "there", thus placing it simultaneously in the process of having been. This temporal process, or internal transcendence, is activated by what Heidegger terms repetition (*Wiederholung),* the sharpest articulation of the imaginative stretching along of *Dasein.*

The character of having been (*die Gewesenheit*) which arises from the future, it should be noted, is to be distinguished from the simple past (*Vergangenheit)* which typifies things present-at-hand (*Vorhandenheit*). The character of *Gewesenheit* implies a vital unity between the living past and the future. This interaction between the past (which is still happening, that is, *das historisch Wirksame* in Eduard Meyer's words) and the future makes temporality the primordial experience of being outside one's self and for one's self. This insight into the temporal structure of human reality offers the philosopher the necessary foundation for ontological speculation, though it comes in the form of the stretching along of *Dasein.* It is the horizon of time that ensures that philosophy and thinking always take the form of a temporally transcendent dialogue. However, the temporally transcendent dialogue that ensues from the Care structure of *Dasein* entails grasping the totality of human reality in the same manner that historical totalities are conceivable from the "own situation" of the historical observer in Troeltsch's logic of history. This historical totalizing, in a matter of speaking, produces a temporally fastened Archimedean point from which to view completeness. This completeness, however, is only conceivable from the position owned by human reality. Grasping the totality in view, as Heidegger suggests in referring to the ecstatic nature of *Dasein*, must mean that the facets of time achieve a unity which is the actual source of totality.

The ecstatic nature of *Dasein* is realized in anticipatory resoluteness *(vorlaufende Entschlossenheit)* and confirms Heidegger's proposition that *Dasein* is historical in its very nature, and not merely world-historical. This historical status or historicizing of *Dasein,* what Heidegger calls G*eschehen* (obviously playing on the cognates of *Geschichte)* challenges the ordinary understanding of history, for the simple reason that the past is denied its privileged position. The process of repetition first brings the equiprimordiality of the facets of time to our attention, and, moreover, suggests the critique at the core of this temporal happening.

By repetition Heidegger intended a temporal process that denies the simple working out of the past in the "today." Repetition entails explicitly handing down possibilities that have been inherited (like death). This brings us to the real theme of history, according to Heidegger, which places *Dasein* in the most auspicious position for an inquiry into Being. Having been freed from the character of the "today" as present (in the moment of vision [*Augenblick*] precipitated by anticipatory resoluteness), history's theme emerges as possibilities which have been existent. The self-referential dynamic of history's theme reveals its existential source. History's theme, then, is really self-inquiry. The source and object of history is experience enduring through time, not isolated facts. The enduring character of experience is expressed in Droysen's principle that history embodies a continual ascent marked by accumulations and additions. Heidegger expressed the interpretative concomitant to this principle in elucidating the self-referential dynamic of the historical happening of *Dasein*. This must mean that history does not merely inform us of what has transpired but of what transpires. What has occurred is the object of history, being-enduring-through-time is history. As *Dasein* is distinguished by not being merely world-historical, but as that entity in which temporalization originates; it stands, so to speak, at the mouth of history. Heidegger has truly supplanted ontology with history, since "what happens" (*Dasein)* is metaphysically anterior to the postulate of an ultimate reality.

## Historicity, Historiography, and Historical Critique

If history as a science, as Heidegger argued, has its roots in the historical nature of *Dasein,* then it follows that the objects of historical research must correspond to the kind of being possessed by *Dasein.* This is a resurrection of Dilthey's insight that historical inquiry is really the rediscovery of the "I" in the "Thou," since Heidegger establishes the historical privilege of *Dasein* through a disclosure of its primordial temporality. These insights led Heidegger to adopt Nietzsche's thematization of history, which he located in the special way *Dasein* happens historically (or temporalizes temporality). Heidegger's use of Nietzsche goes beyond his adaptation of the anatomy of history, since the historical happening of *Dasein* also implies a critique in the special manner that the "now" is deprived of its character as the standing moment.

Nietzsche's organization of history into the antiquarian, the monumental and the critical modes, according to Heidegger, are already present in the ecstatic-horizonal unity of *Dasein*. Heidegger has located these historical forms of apprehension in *Dasein*, but they already had a history in the long-standing debates over the proper way to practice history. Hegel had earlier outlined his views on the subject in his "Lectures on the Philosophy of History" wherein the ultimate historical apprehension is gained in philosophical history, a method that fused the *a priori* with the transitory. Original history, and the other forms of reflective history, in Hegel's outline, are inferior to the apprehension of historical totality afforded by universal history, the final expression of philosophical history. This Hegelian position eliminates the historical efficacy originating in the critical attention of the historical observer. It, however, accounts for the sense of continuity demanded by effective-history. The continuity of universal history is guaranteed by extra-historical forces, while the continuity of effective-history stems from the standpoint of the historical observer.

Heidegger favors the historical continuity that swells up from within historical consciousness. His characterization of the ecstatic temporality of *Dasein* directly speaks to this. As we have already seen in the process Heidegger denotes as repetition, *Dasein* can make possibilities its own through preserving that which has-been-there. This form of temporalization entails both the monumental and the antiquarian in history, since through repetition *Dasein* resolutely chooses possibilities (the monumental) and, at the same time, hypostatizes the existence of that which has-been-there (the antiquarian). But, most importantly, the "today" is deprived of its character as present, since repetition hinges on understanding a possibility which is explicitly handed down from the future. In choosing possibilities that come to *Dasein* in an essentially futural manner, existence becomes detached from the present, thus providing the basis for critique. This implicit Nietzschean transvaluation of values accomplishes a release from what Heidegger refers to as the fallenness of the "they" (*das Man)* and reveals that the historicity of *Dasein* is also the source of authentic existence (*eigentlich Existenz*). Inauthentic existence, or falling prey to the herd mentality, is overcome through the critical posture that can only come through an evaluation of the present from projecting into the future. This imaginative stance can only mean that the self in Heidegger's terms is an ideal. An ideal, however, that is not already formed but one that requires a leap. To relax into the inauthentic mode of existence is to reduce the self to a fact testifying to the absoluteness of the standing moment. It becomes clear that Hegel's wedding of philosophy and history, and the corollary of the absoluteness and necessity of every moment, breeds the herd mentality that Nietzsche and Heidegger argued so adamantly against. The historicization of thought and experience in Nietzsche, Heidegger, and to a lesser extent Dilthey, overcomes history from within history, not from the vantage point of cosmic teleological completion as in Hegel. History is transcended in historicism, but it remains tethered to the ecstatic moment, in Heidegger's terminology. Historicism is predicated on self-transcendence, not the transcendence propagated by a completed revelation.

In disclosing the existential source of historical inquiry, Heidegger initiated a compelling focus for contemporary philosophy. He demonstrated that ontological speculation corresponds to empirical discovery (about history and its methods). This premise means that the object of the historical sciences presupposes the hermeneutic situation of *Dasein.* Heidegger's and Nietzsche's analysis of history exalted the evaluative posture of human existence. They both found the original impulse for this critical stance in the facets of historical inquiry and time. Most importantly, the detachment from the present, implicit in the unity of the facets of time, is actually the historical articulation of authentic existence, that is, existence that is never satisfied with the canonization of the moment. The evaluative posture established from the historicist point of view is a more precise and existential approach to the activist position spelled out in Tillich's assessment of the Western metaphysical tradition in *The Interpretation of History.*

If the historical sciences presuppose an interpretation of the historical existence of *Dasein*, then we can expect that doing philosophy is inseparable from doing history. In this regard, Heidegger took his lead from Count Yorck von Wartenburg, Dilthey's friend and correspondent, who declared that "behavior and historicity are like breathing and atmospheric pressure," and "not to historicize philosophy seems to me, in relation to methodology, to be a carry over from metaphysics." On these terms, Yorck recognized and Heidegger agreed, that Ranke's contribution to the historical school was dubious, since he promoted objective reconstruction (embodied in his methodological dictum *wie es eigentlich gewesen*), and obscured what was living in history–the critique inherent to temporally mediated consciousness. It is this fundamental historical critique that must dictate that absolutism and essentialism in philosophy are fastened to the temporally mediated posture of human existence. Ranke's methodological contributions, apparently, have neglected the interpretative dimensions of history in favor of the facts of history. This approach is predicated on the belief that historical facts possess a discrete spatial status and are never mixed with the present posture of the historical observer. Droysen was more sensitive to the self-critical and self-contributing dimensions of historical investigation; a reception of the bare facts in history must mean that history, or the past, has stopped happening. This latter point is antithetical to the basic premise of effective-history which transformed the past as once-happening into the living past which perpetually instigates historical interest.

## Historical Reason as Critical Philosophy

If we look to Troeltsch, Meinecke, or Heidegger, we can discern the profound consequences of effective-history on philosophy and the interpretation of consciousness. The methodological concerns arising from the early foundations of the German Historical School led inevitably to the ontological reciprocity of philosophy and history. Heidegger's call to resuscitate the question of Being embodied the reciprocity of philosophy and history, since the so-called *Seinsvergessen* of the Western tradition meant that philosophy's steps were to be recovered and rethought. This ambitious project was never carried out by Heidegger, but the structure of *Dasein* revealed that any philosophical speculation is tied to the critical stance of consciousness in time. This anchoring of consciousness eludes the relativism sometimes confounded with historicism, since it fosters the orientation of thought in the horizon of time.

The historical constitution of *Dasein* stems from the original anatomy of historical methodology proposed by Nietzsche and nineteenth-century German historiography. As methodology became more and more identified with worldview and the natural historical outlook of human existence, history became inseparable from philosophy. The identity of the historical and philosophical world is solidified by the moral impulse effected by the self-transcending phenomenon present in Nietzsche's transvaluation of values. The undeniable conclusion of the historicizing of all thought and experience, then, is that epistemology or systematic inquiry of any kind must begin with the life-situation of the historical thinker. Effective-history and the critique at the heart of historicism provide the necessary temporal stream upon which thought is really mediated; however, it does insist on the contemporaneity of the historical thinker as the point of departure of all thinking.

To do philosophy under the rubric of effective-history is to think about history, and this involves the essential unity of the three modes of time. This historical position is analogous to possessing eternity, but always from a posture marked by individual fate. Although the past and the future are critically tied to the present, as demonstrated in Heidegger's analysis of *Geschehen,* he did not intend that any necessary historical content follow from this historical consciousness, hence his clear emphasis on the appropriation of possibilities. Troeltsch, on the other hand, implied the necessary ethical consequences of the historicist worldview. Here, one must acknowledge the general *caveat* of not proceeding to prescriptive measures or content from the formal realm of ontology. Heidegger is indeed emphatic on this point, but Troeltsch, who anticipated a material philosophy of history, remained tied to the Hegelian aspiration of grasping the course of universal history (though from the individuating circumstances of the "own situation" of the historical spectator). In both cases, however, universal history is denied, while the universality of history is affirmed.

Historicity, in Heidegger's work, has become a formal category, and its special significance is intertwined with previous advances in historical methodology. Although Heidegger insists on preserving the unadulterated ontological status of historicity, that is, not jumping to prescriptive conclusions, the critique at the heart of historical consciousness suggests to some extent that a judgment of a good or bad criticism is involved. This insight into the prescriptive layer of Heidegger's ontology seems unavoidable. How, it should be asked, can ontology remain strictly formal after it is historicized and placed at the seat of consciousness? If ontology finds expression in historical matter, then the impulse toward possibility is derived from the descriptive ontology of *Dasein*. Heidegger would also insist that his use of the word Care (*Sorge)*, as the basic structure of *Dasein,* is equally free of prescriptive character. However great Heidegger's admonition against prescriptive advances are, the historical (ontological) grounding of human existence goes beyond the descriptive disclosure of intellectual or psychological faculties. Certainly the historiographical underpinnings of Heidegger's speculative history assures that the essence of history goes beyond description of historical reality. The idealistic tradition that informs Heidegger's ontology had likewise informed the earlier innovations in historiography and this common derivation demands that history go beyond the impartial articulation of historical facts.

If we trace Heidegger's ontology (existential analytic) to his early concerns with historical methodology and the special qualitative character of time in the science of history, then it becomes evident that the value-related connection with which the historian is endowed has emerged as the critical stance of *Dasein* in time. Furthermore, the historical dynamic expressed in the ability of Dasein to appropriate its totality, even when it is still outstanding, is analogous to the historiographical tenet that history can only be apprehended in the dialectic between a particular and a developing whole. This comparison is borne out in Gadamer's assessment that the hermeneutic situation is really the historical drama of human existence. The suppressed ethical dimension in Heidegger's early thought was made even more remote with his foray into National Socialism. However, the predominance of the Care structure in *Being and Time* and the consequent immediacy of a world of concern suggests an essential nature to which we are thrown, in Heidegger's words. If Heidegger has conceived our existence as a sort of project, an end to which we are thrown, then he has at least entertained the idea that there is an end of human existence. One could then weave a moral philosophy, as Aristotle did, on the basis of a *telos* that is imprinted onto our existence.

Ernst Troeltsch saw clearly that the logic of history, expressed in the interaction of the present with future formation, entailed an activism. It seems that Heidegger could not have avoided the same conclusion, if he insisted on the equiprimordiality of the past, present and future, and agreed implicitly with Kierkegaard that life is really lived forward.

This conclusion advances Dilthey's appeal for a critique of historical reason. Critical philosophy, which receives its primary impetus from the ontological character of time (history), then, must transform Kant's unity of apperception into a historical unity which relates a story (narrative). The scope of the narrative might vary, from personal biography to universal history; but, in any event, the universality of history is confirmed as the light of philosophical speculation and developmental contexts must mark the boundaries of all thinking. And, as all thought must begin with self-inquiry, the isolation of introspection will always entail historical critique.

Thus Kant's revolution in metaphysics was unfinished. Indeed, he revealed the contradictions implicit in certain traditional metaphysical concepts, but he quickly restored them to their former absolute elevations in his critique of practical philosophy. Traditional notions of God, immortality and freedom were retained in order to make systematic inquiry intelligible. But, the autonomy of morality, which Kant goes to great pains to establish, is ultimately sacrificed at the bench of universality and necessity. The universal legislative process demanded by the categorical imperative obfuscates the real dichotomy between the world of natural causality and the causality of freedom.

The congruity of the will and pure reason, on the surface, is a remarkable feat in Kant's philosophy. Kant wished to guarantee freedom of the will from sensuous necessity (of the phenomenal world), only to chain the will to the empty dictates of pure reason. It seemed the only way to prove that the will could transcend empirical necessity was to make it partners with pure reason and thus the unconditioned.

The freedom which historicism guarantees is spontaneity and the ability to produce new causal nexus. The facets of time in Heidegger's rendering of the ecstatic constitution of *Dasein* pragmatically proves freedom, and so the future as the condition of all possible experience includes the sensuous in metaphysics. History, as the source of freedom, remains a part of the Kantian revolution, since it too leads to the presuppositions of experience. But, as the ecstatic medium for consciousness in Heidegger's histo-ontology, history not only redeems Kant's causality of freedom, it has transformed transcendental philosophy. Consequently, human reality expresses the freedom of historicism in that it is not subject to historical matter, but gives matter to history. This self-contributing facet of human reality means that the essence of history is the model for philosophical investigations, since the dynamism of historical formation originates in the tension between the particular and the whole.

The attention the relationship between the part and the whole received in historiographical circles is at the core of philosophical history. The manner in which historians addressed this special concern is reminiscent of Hegel's invention in history, but from the historiographical point of view historical seeing makes the interaction between the part and the whole a living, vital complex, not a formal logical construction.

# 3

# History and Ontology

## Historical Recollection and Ontological Nostalgia

In the second volume to *The Philosophy of Symbolic Forms,* Ernst Cassirer proposed a vital connection between philosophy and mythology. In the opening pages of *Mythical Thought,* Cassirer outlines the similarity between the mythical and the philosophical approach to life on the basis of ancient Greek philosophy, which purportedly supplanted the mythological worldview. The connection, he argued, turns on the concern with the problem of origins.[1] The concept of the *arche* is obviously crucial to cosmogonic tales as it is to pre-Socratic cosmology, and even to Plato's rendering of the *eidos*, though Plato struggled to free his thought of any mythological grounding. The domain between the mythological and the philosophical interests us here, as speculation on the *arche* involves the critical temporal consciousness that Cassirer, through his Kantian leanings, claimed as the categorial manner in which we are oriented to experience.

On the one hand, most theories of mythology refer to the origins of institutions, behaviors, and the like, in a primordial or archeaic time that corresponds to the atemporal foundation of philosophical speculation. On the other hand, these creation accounts have come down to us in narrative form, the fundamental structure of historical discourse.

Examined in the light of the structure of mythology, ontology and history appear contiguous. As a proponent of the theory of myth set forth here, Mircea Eliade, following Bronislaw Malinowski, lucidly made the case that history is in fact derivative of the archeaic mythological mindset. Cassirer maintained a similar belief. While Eliade has explicitly expressed the mythological foundation of the modern historiographic mindset, Cassirer acknowledged that mythological and religious consciousness, since it first achieves its awareness in origins and the history of Gods, in contrast to a mere conceptual concern with static Being, can only be conceived in terms of action. As such these forms of consciousness rest on the intuition of the temporal. Time, Cassirer concluded, is the original form of spiritual justification in which historical distance is overcome in an effort to countenance the existence of prevailing institutions through their link with the genesis of the sacred in the primordial, mythical past.[2]

History, which relies on the distinctions between the facets of time, is linked to the timelessness of mythological consciousness, both authors suggest, through the ecstatic continuum which is generated by both forms of historical retrieval. In each case a kind of eternity is formed. That is, an eternalizing movement is realized in which normal irreversible duration is overcome and time remains essentially identical within itself.[3] The identity between the past, as conceived in the genesis of institutions, and the present, which transposes the spirit of these origins into a living presence, constitutes a duration that is integral to the customary delineation of the facets of time. The relation that transpires in the continual reconstitution of primordial events in mythological consciousness contains temporal delineations, but it is concealed in the continuity between the mythical "now," which is so laden with the past. Cassirer echoed Heidegger's critique of the so-called "now" of the "ocular" historians, who resolved the present into a discrete and simple temporal entity suitable for framing, when he invoked Leibniz' characterization of the present as "*chargé du passé et gros de l'avenir*." [4]

The nostalgia for origins that typifies the eternal returning in the ritualistic and narrative retrieval of the past is really an eternalizing return in which the facets of time are pragmatically fused. Mythic consciousness is compelled to forever return to the past to pragmatically ground the institution of civilization (life) itself. Both forms of recollection aspire to a participation in Being. Mythic consciousness repeatedly seeks paradigmatic models initiated in primordial time, and historiographical retrieval is inspired by the effects the living past continue to have on any given present. In both cases events can not be construed as unique and only once-happening. This perpetual returning is a movement that constitutes an eternalizing effect which complements the nostalgia for the *arche*, a longing, so to speak, to participate in Being.

The point at which history and mythology intersect, Eliade's reasoning suggests, is in the tales of ancient demigods who had taken the place of original creation Gods who had abandoned humanity. The creator Gods, known as *deus otiosus,* upon creating the world, had abandoned humankind to their own lot. The *deus otiosus*, as an instance of the death of God perspective, is the beginning of a shift from concern with what the gods had created in primordial time, to a concern with the civil and social structure, that is, the products of human invention. Eliade further noted that civilization flourished in cultures where the gods had either been forgotten or eclipsed.[5] The loss of these gods, then, heralds a history wherein human beings become responsible for instituting and sustaining civil and cultural structures. Human creation, rather than the creation of the gods, became the primary cultural focus, when the creator gods distanced themselves from humanity. Consequently, more compatible divinities (heroic figures), who became more involved with the lives of people, took the place of the forgotten *deus otiosus*. This interaction between the gods and humanity shifted attention away from the original cosmogonic myths, to a more living relationship between humanity and supernatural beings. Naturally, this meant that the focus of these accounts turned to what passed between the gods and humanity, not what had been delivered over to humanity in primordial time.

The assimilation of the religious essence into the human essence, as prescribed in the Marxian transformational critique, is evidenced in Eliade's structural assessment of the archeaic mindset. The primordial time, within which the origin of institutions, and all forms of behavior occur, is supplanted by historical time within which the same creations are made possible. In this substitution there is a clear shift from (pre-)ontological speculation to historical understanding; without the loss of conceptual commitment to the essential structures of existence. These structures become accessible through an understanding of the social nexus implicit in all human invention, not by assigning responsibility for these structures to an extra-historical agent, or essentialist foundations beyond the purview of human imagination.

Another example of the interaction between the gods and humanity that signalled the shift from ontology to history is reflected in the story of gods who suffered death at the hands of humankind. First, these gods always appeared on earth after Creation and were always the victims of human misunderstanding. Through a series of events involving interaction between these gods and humanity, the gods are slain; but before they die they reveal secrets related to the primordial drama of creation. These stories also have the basic structure of all myths, according to Eliade. From the structural point of view they are origin myths, but the mythical time typical of cosmogonic myths has been transformed. No longer is primordial or cosmogonic time involved because the "essential" is no longer tied to ontology, or how the world came into being. The creation of the *arche*, or the institution of the models for all behaviors and entities comprising the world, are the result of an exchange between supernatural beings and a culture's ancestors. The "essential," Eliade concludes, is tied to a *history*. The reason is that the primary focus, now, was on what happened to the gods, not on what they had created. Furthermore, the interaction between humanity and the gods, at this point, meant that these dramatic events were not simply recovered through ritual, but through an act of thought—the telling of a story.

Eliade seemed certain that the process of demythicization begins in the substitution of returning to the beginning with an act of thought, rather than through mere ritual means.[6] This effort to disclose beginnings is common to both archeaic mythology and ancient philosophical speculation. Even as the mythical was systematically being eclipsed by the critical turn to conceptual foundations, and the origin conceived in terms of rational principles in the ancient world, Cassirer reminds us that this was not a real turning away from myth. The movement, he argued, was an assimilation that transposed the power of myth into conceptual philosophy.[7] Cassirer cites the Sophists' use of myth to popularize their brand of philosophy, but the most blatant and ironic assimilation of myth has to be in Plato's philosophy, the conceptual philosophy *par excellence.*

There is a superficial distancing from the mythological world with the introduction of historical consciousness in the Judaeo-Christian tradition of sacred or redemptive history. The prophetic voice of the Hebrew Scriptures focuses on the future and the historical redemption of humanity, what amounts to a turn from concern with origins to an eschatological vision of things to come. The theoretical reciprocity between the Hebrew Scriptures, especially the creation account in Genesis, and the good news in the New Testament, especially the Gospel according to John (In the beginning was the Word), however, retains the temporal dialectic of mythological consciousness that vitally links the past (the time of creation and origins) to the present, in a living continuum that sustains belief in institutive acts and sweeps consciousness into the future.

Even in its truly secular form in the nineteenth century, this historical mind set still contains aspects of mythical thought. (Marx's theory, for example, does not abandon the eschatological outlook of religious soteriologies that anticipate new beginnings.) The conclusion that Eliade so aptly demonstrated is that, today, mythical thought survives most vividly in historiography.[8] That is, historiography is the modern expression of the mythological effort to preserve memory, especially of paradigmatic events and exemplary behaviors. The historiographic recollection of which he spoke was really the way in which one historically enters into oneself. Though mythical consciousness sought solidarity with a "vanished people" in a primordial time, in an effort to commune with creative origins, modern historiographic recollection, following Heidegger, sounds the depths of historical Being in order to create and fulfill possibilities that are enlivened through a recollection of the history of Being, which is aimed at disclosing the origin of historicity *per* se. While universal history, after Hegel, would aspire to the logically precise recollection of historical events and their future consequences, Heidegger's rendering of historical recollection and repetition is more in line with mythical consciousness, since it discloses the primordial temporality and ecstatic eternalizing that grounds human invention and human possibilities.

In living the myth, or in the act of recovering the past, (like the theoretical feats in psychoanalysis), time is overcome and the union between the origin of something (a traumatic event) and a contemporary situation represents a cure for "the pain of existence in Time."[9] It is this vital link between a remote past (which is recalled through ritual and narrative) and a present that precludes the reduction of myth to the form of a story or mere symbolic representation. A story may initiate memory retrieval, but the active historical forces that enliven the mythic signify that whatever reality is revived is to some extent reinstituted and continued. In this state, Eric Dardel has argued, the "mythic man" is thrown from his ontological ground and unable to distinguish the customary facets of time from one another.[10] Time becomes undifferentiated in the repetitive vestigial presence of an originating reality, and the "mythic man" experiences a present which is vitally linked to the past, and produced once again in order to ensure the continuity of existence. Malinowski also emphasized the pragmatic facet of repetition that is embodied in the living faith that originating reality still provokes, but it is not the case that "mythic man" is temporally disoriented when the lived reality of myth is activated. Dardel might just as well have concluded that, though time has the "perennial" character that Maurice Leenhardt spoke of before him, the ontological foundation for "mythic man" and the historical consciousness therein is a living duration that always depends on the vestigial powers of the (living) past, which provide the "pragmatic charter" (Malinowski) for future orientation. "Mythic man" is not lost in the ecstatic resurrection of originating reality, he discovers the temporal circumspection that makes history possible. What becomes a continuation of an originating reality is a repetition of the "now" that "once upon a time" discretely enclosed that foundational reality. Repetition, then, encircles the "mythic man" in a series of moments that constitutes a temporal transcendence, much like Heidegger's conception of *Wiederholung*.

In both instances, the past is deprived of its status as the once-happening, since it is continually reconstituted from the present. Once incarnated in the present, the future, in the form of possibility retrieved from what has been, is simultaneously enjoined. This eternalizing movement embodies the *aufgehobenes* dynamic of Hegel's historical logic, insofar as the past is equally negated, preserved and transcended. Unlike Hegel's logic, however, it is spontaneously enlivened from within the primordial temporality of human existence. Historiographic recollection or repetition, whatever form it takes, though, throws human existence out of the historical moment and into the depths of enduring temporal structures that categorially constitute paradigmatic cultural and historical patterns.

Eliade, it seems, has measured the value of history to mythology in relation to the more practical pursuit of historiography. The fundamental concern of historiography, the various theories notwithstanding, is to preserve the past in narrative form from the vantage point of a contemporary situation. The Renaissance, for example, was a pivotal period in the preservation of the past, though this desire had been at least alive since Herodotus. Eliade acknowledged that the Renaissance preoccupation with the "perfect man" expressed the ambition to seek out exemplary models for life, following the pattern of mythical consciousness in which institutive exemplars are resuscitated through word and rite. This exercise is an example of what he called *historiographic anamnesis*, his interpretation of historical recollection whose goal is to continually reinforce behaviors and institutions by invoking the magical and/or explanatory power of primordial reality and aligning it with present reality. The Renaissance was marked by the specific habit of resurrecting ancient literary and cultural forms. This tendency has expanded since. Through historiographic anamnesis, union can be sought with an array of historical life forms, but more to the point, this form of anamnesis has turned in upon itself. As Heidegger has outlined, and Eliade agrees, human existence has become "possessed of its own historicity." Like the recollection of mythical events, the modern expression of historical recollection can "project man out of his 'historical moment'." This new form of anamnesis, obviously derivative of Plato's epistemology which entails the immortality of the soul, suggests a similar union with a previous historical structure. Equally, as we have noted, the eternalizing function of the temporal transcendence typical of historiographic anamnesis implies an historical immortality that is enlivened by mythical consciousness. Dilthey's sentiment that historical understanding is really the rediscovery of the "I" in the "Thou" is precisely the self-reflexive mechanism intended in the profound way historicity constitutes human existence. In addition, human reality is not merely open to the vastness of the past, it has access to the future which is simultaneously posited. Eric Dardel has insisted that "mythic man" loses his temporal footing and consequently is unable to distinguish "what was" and "what will be" from what constitutes the present. I have construed the vital intermingling of the facets of time in this movement as an active eternalizing that is more reminiscent of Augustine's appeal to still the mind in order to witness how eternity, in which the facets of time are absent, really determines what was, what will be, and what makes up the present (Augustine, *Confessions*, Book XI,11).

Augustine's meditation on time was the prototypical attempt to correct the spatial construal of ordinary time (*Confessions*, Book XI, 20). The parallel to the eternalizing movement internal to the mythic structure is striking. Long before the rise of theories on myth and the emergence of historicism, Augustine had already questioned the actuality of the past and the future which really derive, he argued, from the mind's attentiveness to the present, which in memory holds the past, and which through expectation entails the future. It is only in these terms that the historiographic anamnesis of "mythic man" and Heidegger's circumspective *Dasein* can assert the existence of the three moments of time. If the past and future are tethered to a present without duration, as Augustine clarifies in his meditation, then we are really speaking of the manner in which the mind is temporally stretched along in the process of historiographic anamnesis. The three moments of time collapse into the horizonal stretching along, which is a calculation of re-collected memory that is simultaneously institutive, constitutive and re-combinative. This process corresponds to the ordinary delineation of time, which really devolves into the ecstatic temporal horizon that grounds historical existence.

Martin Heidegger's appeal for the recovery of Being in the Western tradition is the same evaluation of memory and forgetfulness that Eliade addresses through historiographic anamnesis. Heidegger clearly emphasized that history is open on all sides, while Eliade was primarily concerned with the resurrection and momentary incarnation of originating events from mythical time. Heidegger's brand of historiographic recollection, *Wiederholung*, means that *Dasein* is always constituted of the future, the opposite configuration of the form of recollection of which Eliade spoke.

This circumspective stance is a radical reversal of ontological nostalgia, since now the world of *Dasein* is really comprised of possibilities, not of essences that have been delivered over to humanity through the activity of a supernatural ground. The *deus otiosus* might be the real origin of the God is dead movement, but in the hands of Nietzsche, Heidegger and Kierkegaard the movement is radicalized, since history has supplanted ontology and recollection has become a forward-looking process. This modified form of recollection, then, projects human reality beyond the historical moment into future possibilities. But, like the "mythic man" of which Leenhardt and Dardel spoke, *Dasein* becomes contemporary with the past and the future in the activation of *Wiederholung*; through which the depths of historical being resonate in a kind of "perennial" time through which human finitude is overcome.

## The Ontological Structure of *Dasein* and Temporality

The conclusion of Division I of *Being and Time* is that the organization of *Dasein* involves the following structure: the Being of *Dasein* means ahead-of-itself-Being-already-in-(the world) as Being alongside.[11] In a word this structure is expressed by the term

Care (*Sorge*). Care, as the primordial state of the Being of *Dasein,* distinguishes *Dasein* from the other entities in the world, those things ready-to-hand and those things present-at-hand. The intrinsic historicity of *Dasein* further distinguishes it from entities that merely occur in time, thus making it more than a historical datum. As historicity is raised to categorial significance in Division II, the historicism characteristic of the self-contributing dialectic evident in historiography is transformed into the self-constituting nature of human reality.

Care was a suitable designation for the being of *Dasein*, since it is constituted of an environment that is involuntarily engaged. Heidegger inherited his conception of Care from Goethe's character in Part two of *Faust.* Like Goethe's Care, which is derived from antique sources, Care in Heidegger denotes an inclusive sense of existence that does not casually delineate between world and consciousness. In a typical existential manner, *Dasein* has also been defined as that entity whose essence is existence. In other words, there is no fixed human nature. Naturally this means that as potentiality-for-being, the totality of *Dasein* appears to always have something out in front, it is never really complete. Nonetheless, the demands of an ontological interpretation of human existence insists on having a whole-in-view. Heidegger's use of the myth about Care to illustrate the pre-ontological character of *Dasein* signifies that history, as the essential conditioning force of *Dasein,* bears a definite relationship with ontology. The title, *Being and Time*, is the first clue that Being is to be understood in relation to time.

Initially the mixing of history and ontology presents a dilemma. This complexity is summed up in the expectation of totality underlying metaphysical speculation and the existential status of human reality, which is always out in front of itself, never statically complete. Heidegger's primary task in Division II of *Being and Time* addresses this problem. To place the whole of *Dasein* into our fore-having, as he puts it, is to see *Dasein* in the light of *Eigentlichkeit* (authenticity) and *Ganzheit* (totality). The manner in which Heidegger continued his analysis led to articulating the structural whole of *Dasein* in terms of temporality. Not temporality in the ordinary sense of occupying moments of time, but temporality from within the inner complexity of the moment that is never disconnected from the past or future. Naturally this means that the being of *Dasein* is grounded in history and further explains how, as historical, it can form history.[12] Consequently, time-reckoning is constitutive of Care, and a further understanding of the being of *Dasein* becomes intelligible only within the horizon of time.

*Being and Time* was a continued reflection on Heidegger's concept of time germane to the historical world. A decade after his *Zeitbegriff* essay, Heidegger had transformed historical time into the foundation of the existential-ontological analytic of Being. Time-consciousness functions as the key to the totality of *Dasein* and, moreover, is a further step in the clarification of Being. An analysis of *Dasein* serves the dual purpose of illuminating the course to Being and revealing the historical as the metaphysical foundation for a hermeneutical science.

The fundamental historical nature of *Dasein* is the condition for the possibility of both *Geschichte* (history as what has happened) and *Historie* (the science).[13] Heidegger's *Zeitbegriff* essay from 1915, in primarily addressing the task of the historian, does not assume the ontological proportions of historicity in *Being and Time*, but it certainly does foreshadow the explicit critique at the core of historical thinking in *Being and Time*. As explicated in the *Zeitbegriff* essay, the task of the historian can serve as an exemplar for history as the theme of a fundamental ontology. History, in *Being and Time*, remained grounded in concrete life, and it is determined equally by the objects of history and the ontological force behind it– *Dasein*. In fact, the two, the object of history and *Dasein*, seem to merge in Heidegger's analysis. To justify this correspondence is to realize that he who inquires into history is he who makes history. Dilthey's formulation that history is always accompanied by self-inquiry is borne out in Heidegger's centering of history in the existence of *Dasein*. At the same time, as Dilthey would have it, historical activity is not simply a self-construction, but a genuine reaction to the effects of another presence.

Like Ernst Troeltsch before him, Heidegger was antagonistic to relativism and sensed in historicity, as an essential condition of *Dasein,* the means to take responsibility for one's "own situation." Being situated in history means that the consciousness of *Dasein* cannot be divorced from the consciousness of the historian. In this context, the purpose of history, always geared toward critique as it is linked to the future, is not a mere fascination with past events or personalities, but a unity of past, present and future directed toward self-possession and self-creation. Along with Troeltsch, Heidegger implied the lively connection between criticism and creation in historicity. The ontological analysis of historicity is strictly a formal enterprise, though it implies a praxis by virtue of its essential critical aspect. In fact, the circumspective concern that emerges from the temporal way that *Dasein* is in the world means that human existence is praxically disposed at all times, except when the theoretical point of view takes over. Heidegger reminds us, then, that theory evolves only with the privation of praxis which is ontologically prior (BT, 409). The positing of the priority of the practical in the Care structure of *Dasein* means that a metaphysics (which clarifies the possibility of experience) is coincident with the awareness of a world. Heidegger uses the expression *Vergegenwärtigun*g for this type of seeing or visualizing, placing emphasis on the its cognate *Gegenwärtigen* which means to make present (BT, 411). This special way of making-present is different from uncovering something that is merely present-at-hand, since the projected state of *Dasein* is always toward potentiality, an implicit state of involvement whose internal structure consists of a temporal interconnectedness comprised of expectation of possibility. Furthermore, the visualization of objects of concern is not simply a means of representation, another connotation of *Vergegenwärtigun*g; it is the primary way that the predisposition toward being involved in the world is disclosed to us.

## Historicity and Being Towards an End

As the key to both *Eigentlichkeit* and *Ganzheit,* Heidegger's analysis of death plays a central role in the historicity of *Dasein.* I would like to recall some of the main components of Heidegger's analysis, insofar as they clarify the issues of authenticity and totality and prepares the way for an understanding of how *Dasein* happens historically.

The distinction between *Eigentlichkei*t and *Uneigentlichkeit* is made clear in Heidegger's analysis of death. Being-toward-possibility takes on a profound dimension in the existential analysis of death. As the ultimate possibility in human existence, death remains an unsurpassable possibility that primarily comes to *Dasein* through anxiety (of nothingness). In being-toward-an-end, death is incorporated into the existence of *Dasein* at the moment of birth, and is therefore a possibility from the very start. Consequently, *Dasein* is immediately anxious, for its birthright is to be always bordering on nothingness. The logic of being-towards-an-end entails the dialectical relationship between being and nothing.

Heidegger intended to convey the incentives disclosed by being-toward-an-end. His insistence on the ever-present possibility of being-toward-an-end underscores the incorporation of the future in the existence of *Dasein.* As necessarily future-oriented, *Dasein* is always comported toward itself. (This self-constituting dynamic is analogous to the self-contributing movement of history that was addressed in Droysen's theory of history.) The acceptance of being-toward-an-end is the key to *Eigentlichkeit* or self-appropriation. As the ultimate possibility for *Dasein,* death signals all the other possibilities that come in advance of the possibility of no longer being-toward-possibility. In turn, this provides a perspective which actively anticipate possibilities. Thus, being-ahead-of-oneself, in resolutely anticipating death, is coming to terms with the historicity of *Dasein.*

Whether it is regarded as the key to *Eigentlichkeit* or *Ganzheit*, Heidegger's analysis of *Sein zum Tode* establishes the individuating tendencies of *Dasein. Sein zum Tode* individualizes, and falling prey to the familiar pattern of holding death at bay leads to the mechanical life of the herd, the *mauvais foi* elucidated by Sartre. Regarding death as something that is always one's own and not to be transferred to or disguised by the "they" (*das Man*), is claiming the existential birthright of always running-ahead-of-oneself. Therefore, any effort to ground such an understanding of being-toward-death in empirical evidence, that is, establishing an exemplar, as Victor Farias does in his controversial book on Heidegger's relationship to National Socialism, construes death as simply an end attached to *Dasein* and outside the individuating movement of being-towards-death.

That the future cannot be disconnected from the other facets of the temporality of *Dasein,* namely the past and the present, implies a unity that is reminiscent of Bergson's time as duration. *Dasein,* on these terms, is the heterogeneous continuity of which Bergson spoke. The past, as intimately connected to the present, is not the past that has transpired, that is, *Vergangenheit*, but the living, vital past—*Gewesenheit.* This function of the active past, in service of a continuous unity or heterogeneity, is similar to Troeltsch's conception of the historical unconscious as effective tradition. Likewise, the future is inextricably attached to the present as constant possibility. Therefore, by virtue of this temporal stance, *Dasein* is constituted of a circumspective vision that amounts to a historically transcendent being—transcendent within the parameters of the horizon of time, not with respect to a realm of essences or a transcendental ego. The very idea that the past and the future are incorporated into the present suggests a practical transcendence. *Dasein* does not find its existence in a succession of moments, its experience of time is an "experience of longing, of pursuit, of restlessness."[14]

## Self-Discovery and Time

H. Stuart Hughes, in *Consciousness and Society* (1958), pronounced that time as duration was one of the important revelations of the decades between 1890 and 1930. Wyndham Lewis's *Time and Western Man* (1927), though it is a diatribe against what he called the cult of time, indicates the passion that surrounded the rise of temporal consciousness. Lewis was somewhat out of his league in denouncing the time philosophy of Whitehead, Bergson, and even Joyce, but he was handicapped by a profound denial of his past and the value of introspection. This phenomenon was by no means restricted to the intellectual environment of Europe. Josiah Royce was motivated by a similar conception of time, and, in arguing for time as the form of practical activity, he has identified with the heritage of Idealism and joins with Heidegger in demonstrating that the past and the future stand in essentially practical relations to the present.[15] Consequently, for both philosophers "the past, the present and the future constitute a single whole, whose parts have no true Being except in their linkage."[16]

Josiah Royce's discussion of the temporal and the eternal in 1901 bears a remarkable resemblance to Heidegger's meditation on time in *Being and Time*. The similarity between the two thinkers is striking and might be partially explained by Royce's experience as a student in Germany in the 1870s. Royce's basic premise, that time is the form of the will, is analogous to Heidegger's claim that time is the most pervasive form of finite experience. Royce's words speak for themselves:

> In pursuing its goals, the Self lives in time. And, to our view every real being in the universe, in so far as it has won union with the ideal; and accordingly, so far as we can see, is living time. Whoever, then, is finite, says 'not yet', and in part seeks his other as involving what, to the seeker, is still future. For the finite world is general, then, as far as human beings, the distinction of the past and future appears to be co-extensive with life and meaning.[17]

The avowal of "not yet" on the part of finite being is much like *Dasein* always having something outstanding. Royce's further clarification of the activity of the Self suggests that it, like *Dasein,* is always something that is mine alone and constitutes a totality: "in so far as I am this individual and no other, i.e., my goal as this Self is still not any one point or experience in my life, nor any stage of my life, but the totality of my individual life viewed as in contrast with the lives of other individuals."[18] The self-constituting dynamic of Royce's Self and Heidegger's *Dasein* reveals an existential context of being situated within a personal temporal domain.

The contrast between the Self and the lives of other individuals distinguishes the Self as a more or less private existence as *Dasein* (*eigentlich Dasein*) is always and in every case one's own. Both Royce and Heidegger identify this personal Self as bound to past and future that upon examination is the "really deep and important persuasion that one ought to possess or to create for oneself."[19] The dynamic of selecting from existence a certain "portion of remembered and expected, of conceived and of intended life" as that of one's own and individual Self, for Royce, is characterized by an "ought to be able."[20] *Dasein,* according to Heidegger, finds itself already immersed in such a temporal-ecstatic dynamic. The consideration that one "ought to be able" to appropriate memory and expectations places Royce more in line with Kierkegaard and Troeltsch who both identified historical existence with the ethical. Royce ultimately concluded that the temporal unity of the Self is an Ethical Category.[21] It remains, though, that Royce's conception of the Self and Heidegger's *Dasein* are compatible views of human existence, for individual existence, in each case, defies subsumption under a genus. And, in the weighing of the past and the future, a "task is here hinted to me as my task, my life-plan" and "by this, and not by the possession of a Soul-Substance, I am defined and created a Self."[22] The consideration of existence within a horizon of time is the manner in which the Self is defined and created, not through the inheritance of a formal pattern or nature to which each private existence must conform. This process implies the necessity of the historical dimension in the pursuit of meaningful existence. Meaning and creation are thus contingent upon the historical realm within which the Self participates in practical-historical transcendence. The very same precept is expressed in Marx's sixth thesis on Feuerbach, that is, "human existence is no abstraction inherent in each single individual. In its reality it is the ensemble of social relations." As the product of concrete relations, this ideal self as conceived by Heidegger and Royce emerges, in the words of John Dewey, when the imagination lays hold of the possibilities offered to thought and action. Dewey also suggested that this ideal self does not receive its authority in "some prior complete embodiment" but is the result of real seeing and sensuous activity wherein "old things" are imagined in new relations that will serve new goals which the goals themselves help to design.[23] Dewey's critique of institutional religion, in *A Common Faith,* is analogous to Marx's and Feuerbach's transformational critique of Hegel's systematic philosophy. Hegel's philosophy, which epitomizes the essentialist premise of the authority of an antecedent reality only destroys, according to Dewey, the dialectic between faith and the object of faith. His critique is relevant to the dynamic of historiographical anamnesis in Eliade and Heidegger, since his reversal of the identification of the religious ideal with an *a priori* Being restored the ideal to the conditions of historical existence.[24] The restoration of the religious ideal to possibilities in existence meant that the authority of the ideal was now to be sought in the imaginative grasp of "old things in new relations serving a new end which the new end aids in creating," not in the encumbrance of a dogmatic completion that is delivered *a priori* to the present. Dewey called the temporal dialectic between the ideal and the actual, "God."[25] More importantly, it is generated by circumspective projection, the creative union of originating reality that is, through repetition, brought to life and reconstituted through the grip that future ideals have upon the present.

The distinction of *Dasein* from everything that is present-at-hand or ready-to-hand leads to its experience of temporality in the phenomenon of *vorlaufende Entschlossenheit.* This running-ahead-of-itself or *Vorlaufen* is the distinctive mode of temporality resulting from the Care structure of *Dasein. Dasein*, then, cannot be defined in terms of a simple soul-substance. Constituted of Care, it only becomes intelligible as an entity that is always-ahead-of-itself. In a manner of speaking, *Dasein*, *qua* Care, is always going out after itself. Heidegger's analysis of being-toward-death has already disclosed both *Eigentlichkeit* and *Uneigentlichkeit* as options for *Dasein*. The activity of anticipation or resolutely engaging one's finite borders is that form of understanding that leads to acquiring design over one's existence.[26] In actively dispersing all fleeing self-concealments,

*Dasein* is brought into the resoluteness of action.[27] *Sein zum Tode,* as typified in *vorlaufende Entschlossenheit* brings *Dasein* face to, face with its individualized potentiality-for-being.[28] Anticipation makes *Dasein* future-oriented. Thus, in having been in advance of itself, *Dasein* is also understood as having-been. Not only does the character of having-been arise from the future, but running into the future initiates action through an immediate concern with one's surroundings.[29]

Temporality, then, is not something that has been assembled in a cumulative manner, it is a swelling up of the unity of the three facets or ecstasies of time as Heidegger calls them. To be precise, he says that *Zeitlichkeit ist das ursprüngliche 'Ausser-sich' an und für sich selbst.*[30]

The temporalizing of temporality, or the manner in which time is unified, is the historicity of *Dasein.* It should not be understood that because of this foundation *Dasein* merely occurs in history. More accurately, historicity serves as the basis for historiographical understanding.[31] The claim that *Dasein* is historical reveals an ontological privilege. Although related to the ontical assertion that *Dasein* occurs in history, it suggests, more importantly, that *Dasei*n is a temporal happening. In this regard, the structure of Care has revealed the significance of ahead-of-itself as the primary facet of time for *Dasein.* One can expect that the connection that historicity has with historiographical understanding will permit the same access to the primordial levels of consciousness as Eliade's historiographic anamnesis.

Heidegger's analysis of *Eigentlichkeit* goes hand in hand with the exposition of the resoluteness derived from *Dasein* continually being-ahead-of-itself. Consequently, there is both the possibility of *eigentlich* temporality and *uneigentlich* temporality. That form of temporality appropriate to *Eigentlichkeit* begins with the future in resoluteness. The present, therefore, is not to be understood as a mere now, wherein something is present-at-hand, but must be understood as an active present, more complex than the static conception of the now. After all, Heidegger calls the facets of time ecstasies, and the *eigentlich* present becomes a "moment of vision" or *Augenblick* within which a view of possibilities and objects of concern are encountered. The *uneigentlich* present, in contrast, represents a mere now that is occupied by something and is characterized by the mood of curiosity in the presence of something-at-hand. Nothing fills the *Augenblick,* in the sense above, it is an ecstatic moment wherein *Dasein* is always ahead-of-itself, grounded in an involvement with things ready-to-hand. The *eigentlich* temporalizing of *Dasein* signifies that time does not come in succession but forms a unity.[32] This ecstatic unity possesses something like a horizon, according to Heidegger.[33] The facets of time do not simply displace one, there is in each case a "to where" (*Wohin*) that is attached to each ecstasy. Heidegger calls this a horizonal scheme wherein the future is always aligned with a for-the-sake-of-itself to which *Dasein* comes toward itself. *Dasein*, already finding itself in a situation (*Befindlichkeit*), encounters the horizon of *Gewesenheit* or having-been. Finally, the present, finding *Dasein* always alongside something, is designated by the "in-order-to" (*Um-zu).*[34] On the basis of this horizonal constitution there also exists a world, and most pointedly, Heidegger reveals his legacy of Idealism when he concludes: *Wenn kein Dasein existiert, ist auch keine Welt 'da'.* [35]

This inner temporal complexity mirrors a self-comportment that hearkens back to the historiographic understanding of the early German Historical School. It is likewise embodied in the creative impulse that typified archeaic mythological consciousness for Mircea Eliade, since *Dasein* is continually involved in its own creation. In other words, the cure for overcoming time, the nightmare of history as Stephen Dedalus bemoaned, is within time-consciousness itself. The creations and origins that now occupy time and the historical are not cosmogonic but self-constituting. Heidegger's solution would be Stephen Dedalus's solution: the past is never over and done with if it can be continuously used over again.

## Temporality and Historicity

Like Troeltsch, Heidegger questioned the enterprise of metaphysics in the traditional sense (as indicating something eternal and extrahistorical). When Heidegger addressed metaphysics he looked to the structure of the finite existence of *Dasein* whose being is understood within the horizon of time. Karl Löwith, a student of Heidegger's, called Heidegger's untraditional brand of metaphysics a "finite metaphysics of finiteness."[36] As an existential construction of history, on the basis of the temporality of *Dasein,* Heidegger's work must be regarded as an attempt to radicalize Dilthey's work. Löwith understood Heidegger to be seeking a redemption of modern historicism, as sought by Dilthey, and tentatively worked out by Troeltsch, through absolutizing historicity itself.[37] Heidegger, in looking to the phenomenon of temporality temporalizing itself, sought a self-redemption of historicism in the same manner that Troeltsch looked to the essence of historical movement for the solution to overcoming historicist pitfalls. The way into Heidegger's philosophy of history is through the concept of *Geschichtlichkeit* or historicity and the peculiar manner in which Dasein is said to happen historically (*Geschehen*).[38]

In pursuing an analysis of the historicity of Dasein in *Being and Time* we should not lose track of Heidegger's *Zeitbegriff* essay. It is instructive to keep in mind in order to measure the enterprise of historiography against the ontological description of *Dasein. Being and Time* makes the case for the ontological depths of the temporality of *Dasein*, which in turn makes the science of history as well as time-reckoning (datability) possible. Heidegger's *Zeitbegriff* essay, however, reveals a preoccupation with the science of history and the task of the historian. Chronologically, Heidegger began with an investigation of the science of history which obviously grounded his subsequent attempt to formulate a fundamental ontology. In 1916 Heidegger saw the "culture creating capacity" of humankind as the objectification of the spirit and stretches of years as mere "fitting posts, but meaningless in themselves."[39] In this essay, as we have seen, Heidegger made the case for value linking the past to the present in the selection of information out of the abundance of available data. The link between past and present (and future) in *Being and Time* becomes centered in *Dasein*. Admittedly, there is an ontological orientation in *Being and Time* and so one must guard against parallels between conclusions drawn on the ontical level and those founded on the ontological level. It remains, in spite of this difference, that *Dasein,* in Heidegger's thinking about history, later occupies the position that the dynamics of value-connectedness in the context of the historian's task once occupied. Moreover, if it was an observable turn that Heidegger made in his work, upon what was he turning or what was he turning away from. The historiographic understanding of the historian, I would argue, is inseparable from the ontological introspection of *Dasein.*

If Heidegger wished to remain consistent in his claim that historicity and the temporal footing of *Dasein* are responsible for the science of history, should we not also accept the conclusion that the ontological structure of historicity entails the value-relatedness inherent to historical inquiry expressed by Rickert and Simmel a generation before? The resolution to this query lies in the special status of *Dasein*, which is at once ontical and ontologically privileged. I would argue that the exemplar of the value-related task of the historian is transferred to the evaluative structure at the foundation of the historicity of *Dasein*.

Heidegger's analysis of *Dasein* suggests a further clue for overcoming historicism. Thus far, we have established, in the absoluteness of the historical nature of human existence, that historicism is not so much a threat to self-knowledge, but an asset. Heidegger's positive evaluation of the hermeneutic circle is a reflection of the benefits of the historical footing of *Dasein*. Historicism in the "bad sense" [40] is transcended because of "the radical and finite subjectivity of resolute decision, a subjectivity which contains no subtle appeals to a divine plan or a quasi-divine dialectic of progress for its justification."[41] In addition, historical relativism is combated through the "absolute, that is, constant structure of human historicity."[42] In light of the absolutizing of history, *Eigentlichkeit*, relative to *Dasein* and the science of history, is not just the result of an antiquarian emphasis on the past but the "disclosure of monumental possibilities for existence as such."[43]

Nietzsche's influence on Heidegger's philosophy of history is evident in this consideration of the disclosure of possibilities. Perhaps Nietzsche's most important contribution to Heidegger's philosophy of history was his innovative consideration of the present as embodying critique. The stretching along of *Dasein* consists of the way in which the future and having-been are merged with the present. In thus looking to the future, as the source of possibilities for existence, the present is denied its aspect of being now, and *Dasein* becomes detached from the *uneigentlich* everyday, public self. As authentic (*eigentlich*), the history which is monumental and antiquarian is necessarily a critique of the present.[44] The *eigentlich* historicity that is the ground for *eigentlich* history is the source of the overcoming of historicism as both determinism and relativism.

Heidegger's appropriation of Nietzsche's delineation of the aspects of history, that is, the antiquarian, the monumental and the critical is significant in understanding the manner in which history can provide the means to overcome the historical (relativism). Heidegger was most concerned with the critical aspect of history. Strictly speaking, this facet of history is the source of the truly historical. The reason for this must be that the critical mode demonstrates that history is not comprised of a mere sequence of nows.

This is evident in Heidegger's appraisal of Ranke, and his followers, as mere antiquarians, connoisseurs and collectors of the past. The critical facet of history opens up the way to historical disclosure. Thus it represents the space, as embodied in the future, wherein Being comes to light. Heidegger laid the foundation for such a consideration of truth in the first part of *Being and Time*, where, through the use of the metaphor of the *Lichtung*, he proposed that the more important aspect of Being is that space which originally allows for the disclosure of Being. This metaphor represents a genuine challenge to traditional metaphysics. Ernesto Grassi, as we have seen, in *Heidegger and the Question of Renaissance Humanism*, suggests that Heidegger has replaced the traditional metaphysical presupposition of the *lumen naturale* with the idea of the *Lichtung*. Accordingly, "it becomes no longer necessary to ask about what exists in the way traditional rational deductive metaphysics does."[45] Furthermore, rational discourse does not adequately capture the presence of the original space of disclosure; rather, the poetic and the metaphorical become the tools for the expression of historical truth. Historicism, considered either as determinism or relativism, is overcome through the relation to one's own historical nature, as a creature facing the future– the space of disclosure.

In looking to *eigentlic*h historicity, as the solution to overcoming both historical determinism and historical relativism, it must be that we should look to the manner in which *Dasein*, as human existence, expresses a connectedness or continuity through time. The aim of this strategy, following Heidegger, is to throw light on the basic integrity of *Dasein* that protects it from the kind of determinism to which entities under the rubric of the natural sciences are subject. It has already been implied in the running-ahead-to-death, that typifies *eigentlich* existence, that there is an effective connection between the present and the anticipated future, which draws possibilities in-between (the present and anticipated future possibilities) to *Dasein.* This effective connection or appropriation functions between the end of *Dasein,* in death, and the end of *Dasein* at birth, to complete the totality of *Dasein. Dasein*, thus being closed off on either end, exists as a form of natural *epoché*. Accordingly, this means that the historical constitution of *Dasein* eliminates prejudices such that the thing itself is disclosed in the way that it is temporally disposed. To characterize the connectedness of life (*Zusammenhang des Lebens*) as a mere sequence of nows is to transgress the essential qualitative nature of time applicable to the historical world. It seems that with such an effective wedding of nows a self-constancy is maintained. The nature of this self-constancy is explained in the particular way that *Dasein* happens historically (*Geschehen*). If the being of *Dasein* is constituted of being-in-advance, then the passage of experiences in between birth and death must already be incorporated in *Dasein.*[46] The question of connectedness is directly related to the way in which *Dasein* is stretched along. This incorporation of *Dasein,* emanating from the depths of its temporal nature, does not merely express continuity, but being part of the nature of *Dasein*, it also explains how the connectedness is taken possession of. This analysis of *Dasein,* how the totality is already present in *Dasein*, explains why human existence can represent the way into Being, that is, *Dasein*, as a potential whole, is the original ontological building block. Ordinary history has not been helpful in disclosing the existential-ontological constitution of historicity, since its emphasis has been on what has seemingly transpired. Indeed, history, for Heidegger, only makes sense as derived from the historicity of *Dasein.* The privileged position bestowed on the past in ordinary history, in being supplanted by the context of becoming (*Werden*), proves, for Heidegger, that anything that has a history is characterized by development (*Entwicklung).*[47] Heidegger wished to eliminate the special status of the past in history in order to reveal how all the facets of time co-operate to make history an active and effective coherence: *Geschichte bedeudet hier einen Ereignis und 'Wirkungszusammenhang', der sich durch 'Vergangenheit', 'Gegenwart', und 'Zukunft' hindurchzieht*.[48]

The developmental character of *Geschehen* was first demonstrated in the manner that *Dasein* resolutely comes back to its own thrown state through anticipating its ownmost possibility (death). Since death, as being-toward-an-end, clarifies the nature of ownness, or that which is inseparably part of the nature or definition of *Dasein*, it represents the point of departure for any further understanding of the quality of ownness. With this special status Heidegger justifiably refers to death as the ownmost possibility for *eigentlich Dasein.*

## Repetition as the Dynamic of Historical Logic

The activity of anticipating death (an activity that Heidegger distinguishes from expectation) makes *Dasein* an entity that is essentially future-oriented. In the process of handing down possibilities, the result of anticipating one's own death, resoluteness becomes a kind of repetition (*Wiederholung*) of the possibility of existence.[49] Heidegger's rendering of repetition clarifies how *Dasein* happens historically and provides a more precise understanding of how Being is disclosed. The light that was the presupposition of traditional metaphysics is transformed into the historically revelatory posture of *Dasein*, which is activated in the process of *Wiederholung.*

*Wiederholung,* as the way of handing down possibilities from the future, places *Dasein* equally in the process of having-been (*gleichursprünglich gewesend ist*).[50] And, as repetition is focused on the future, it does not have the ordinary sense of activating something that has transpired. Rather, it signifies a union between a future possibility which has-been-there through resoluteness. It is made in the *eigentlich* present, *Augenblick*, which in turn causes disinterest with the simple working out of the past and whatever might be gained by the future. Consequently, history for *Dasein* does not have its main importance in either the past or the connection the now has with the past, but in *Geschehen* which springs from the future.[51] The way in which *Dasein* happens historically is grounded in *Dasein* being essentially open. This must mean that it does not have a particular nature, but, more accurately, has a history.[52] The *Geschehen* which lies in *vorlaufende Entschlossenheit* and *Wiederholung* Heidegger designates as *eigentlich Geschichtlichkeit*.[53]

Repetition, since it is at the heart of *eigentlich Geschichtlichkeit*, requires further scrutiny, for here Heidegger has put a microscope to the historical movement typical of *Dasein.* Like Kierkegaard before him, Heidegger has located history within the individual, and so movement must characterize its inner logic. Likewise, Kierkegaard, in his ambition to counter Hegel's system, was intent on developing a metaphysics of interest, one that derived from the inward movement of life. It was his concept of repetition that was at the center of this metaphysics, and it is related to the critical mode of history that Heidegger inherited from Nietzsche.

As a technical term, *Wiederholung,* as we have seen, is removed from its ordinary meaning, like so many of the common terms Heidegger employs. However, it was similarly employed by Kierkegaard in his essay on experimental psychology eighty years before Heidegger. The fact that Kierkegaard's philosophizing has clearer connections with life (his own for the most part) makes the concept of *Wiederholung* immediately intelligible. That Kierkegaard's analysis of repetition serves to demarcate ancient (essentialist) philosophy from modern philosophy, typified by Leibnizian monadology, is also illuminating.[54]

Kierkegaard endowed repetition with the same movement as recollection, only the movement occurs in the opposite direction.[55] Seemingly, repetition is somewhat of a paradox if based on the function of recollection, which casts our attention backwards into the past, and onto essences that are anterior to history. The decisive thrust of Plato's theory of *anamnesis*, which Kierkegaard had in mind, is that life originates in ideas or forms that are recollected. To recollect forwards is precisely what we are to understand by Heidegger's concept of *Wiederholung,* since it is grounded in resoluteness running-ahead of *Dasein.* The disclosing of a possibility and the return to a resolute *Dasein*, set on action, is the essential movement of repetition. It is not a reconstitution of a by-gone past but the disclosure of possibilities upon which existence can be modeled.[56] Recollection forwards reverses the ordinary function of modeling. Life is not merely open to possibilities but is bound by them. Though the ordinary function of modeling is reversed, it suits the nature of *Dasein* that, in Kierkegaard's words, must live life forwards. If recollection has been the interest of traditional metaphysics, then Kierkegaard's admonition that it is also the interest upon which metaphysics founders indicates that repetition's movement must first be directed inward and then forwards.

The course of repetition for Kierkegaard was the road to happiness, unless one chooses not to live but at the "moment of birth tried to find a pretext for stealing out of life, alleging, for example, that he has forgotten something."[57] Or, as spelled out in Heidegger, repetition is subverted if *Dasein* falls prey to the herd mentality and succumbs to everydayness. In both cases we can see the fleeing from life by which one is constantly tempted. This fleeing, however, in a privative way, places *eigentlich* existence in relief and therefore serves an invaluable function. Kierkegaard stated his case in reference to ancient philosophy which established the world's possibilities as a *fait accompli*, and Heidegger frames *uneigentlich* existence in terms of mechanistic modeling to which one is forever vulnerable. The movement of repetition, for Heidegger, shuns the traditional primacy of the past in history, it does not necessarily strive for progress, and it revels in the moment of vision that is brought to light in *Dasein* handing down possibilities to its own facticity. Likewise, Kierkegaard's repetition is not marked by the hope of adventure and discovery (progress), nor is it burdened by the sadness of recollection (the past); "it has blessed the certainty of the instant." [58]

The movement of repetition is clarified in Kierkegaard's allusion to ancient philosophy. The claim that all knowledge is recollection must affirm that everything has already been. But repetition in its forward-looking form "affirms that existence which has been now becomes."[59] Kierkegaard dramatically presents his case through the example of a forlorn lover who, in Kierkegaard's eyes, typifies the ancient worldview of recollection. This example is no doubt autobiographical, and so Kierkegaard's own sense that he was an exception to the normal human being must partially explain his intentions. The example of the torn lover reveals that the lover is mostly driven to poetry and excess melancholy in his love for a young woman. This meant, according to Kierkegaard, that the young woman was not the real object of the young man's love, but that the idea of being in love had possessed him. Moreover, the young man's fever over dreading the loss of love, as the account goes, is real– recollection (possession by the idea) being predicated on loss.[60] The young man's mistake was incurable "and his mistake was that he stood at the end instead of the beginning."[61] Analogously, we can regard Hegel as suffering from the same malaise of historical recollection. Possessed by the Idea, Hegel also stood at the end (of time), and his philosophy of history is only conceivable from this completed point. As in the case of the young lover, where real life is reduced to melancholy over the loss of love, Hegel was consumed by the idolatry of the Idea.

For all of Kierkegaard's experiment with the psychology of repetition it remained hopelessly transcendent and beyond his grasp. Not for Heidegger, however, who shows no modesty in wrestling with the ontological fabric of existence. Kierkegaard admitted that he failed at the point where he could not erect himself upon himself, finding it impossible to circumnavigate his own existence. [62] Heidegger, seemingly, found an Archeimedean point in death, although it is clearly not to be conceived as a mere point. Furthermore, the self-reflection and self-possession of historicity is the self's manner of building itself upon itself. The privileged status of *Dasein*, as the clearing through which Being is manifested, has made Heidegger's venture into the existential-ontological realm more manageable. To describe human existence in such historical terms is to assert that

*Dasein* is always comported toward creating something out of that which already constitutes it.

It remains that Kierkegaard's investigation of repetition clarifies that Heidegger's concept of *Wiederholung* is an existential twist to Plato's theory of recollection. Repetition is part and parcel of the dynamic of *Geschehen* and is grounded in *Dase*in being ecstatically open, neither dominated by the past nor set on progress, but always open to its own history.[63]

Still, of what does the connectedness of *eigentlich Geschehen* consist? Is it a series of resolutions that must be further connected or do they already form a unity in the special way that *Dasein* happens historically? These are some of the questions that remain at the conclusion of section 75 in *Being and Time.* But hasn't Heidegger already referred to the connectedness of *Dasein* in the activity of repetition, that, although essentially future-oriented, equally places *Dasein* in the process of having-been and in the *Augenblick*, whereupon *Dasein* takes over its own thrownness? This most certainly demonstrates the ecstatic openness of *Dasein*, but is this proof of the connectedness of which Heidegger speaks?

Heidegger did not intend to ask how resolutions or experiences are subsequently connected but how they are inherently connected given the *eigentlich* historicity of *Dasein*. The unity of a sequence of experiences is always there for *Dasein,* but *Dasein* constantly faces the threat of distraction from the "they," which is most eloquently revealed in the fleeing in the face of one's own death. Losing oneself in the "they," then, is the opportunity to recognize the inherent connectedness of *Dasein* and to subsequently pull oneself together, so to speak, in the connectedness that is already there: *Die Entschlossenheit des Selbst gegen die Unständigkeit der Zerstreuung ist in sich selbst die erstreckte Ständigkeit.*[64] This continuity, or permanence, is already the inclusion of birth and death and the in-between such that *Dasein*, in this state of incorporation, finds itself in the instant of the moment face to face with its world-historical situation, that is, to what is present at hand.[65] The ecstatic dialectic of the historicity of *Dasein* is thus a kind of temporal gyroscope, and resolution makes up the loyalty of existence to its own self.[66] It does not simply last as long as an act of resolution lasts, but persists in every moment of vision that may spring from the act of resolution. The connectedness of *Dasein* is not the result of joining moments one to the other, the temporality of *Dasein* has already been stretched along.

*Uneigentlich* historicity, on the other hand, conceals the temporal-ecstatic-horizon, and *Dasein* is geared only toward what is present-at-hand, that is, the present is realized on the basis of what is actually left over from the past, like the remains of a partially consumed meal. Engaged with only what is world-historical, the chronicled data of history, *Dasein* becomes disloyal to its own existence and is lost in the ossification of the present. The internal historical dynamic forever comported toward creating something from what already constitutes *Dasein* is covered up. *Eigentlich* temporality, on the other hand, in the *Augenblick*, actively running-ahead and geared toward repetition, denies the 'now' status of the present and frees one from the usual fare of the "they"(*Üblichkeiten)*.[67] The loyalty to one's own existence is comprised of the openness to possibilities that are realized through resolute repetition. Disloyalty to one's existence is a form of blindness in the face of one's possibilities and equally results in viewing the legacy of the past as mere artifact. In the first case, history generates an activism toward one's own existence; in the latter case, history becomes a matter of aesthetic appreciation and archeaeological consumption. Repetition, understood as the incorporation of the process of having-been, in resolutely running-ahead-of-oneself, organizes *Dasein* in an ecstatic-temporal-horizonal nexus.

Historiography, in turn, if it is to be genuine, must take its departure from the historicity of *Dasein.* Heidegger recognized this connection in Nietzsche's delineation of the threefold character of history as a science. Nietzsche, however, failed to articulate the unity of the threefold character of history.[68]

We have already seen the function of monumental history in Heidegger's analysis of historicity. The identity with a hero, however, plays itself out in a futuristic manner in resolutely disclosing possibilities and coming back in repetition to the bold "there" of *Dasein*. In making the possibility one's own through repetition, *Dasein*, when geared toward the future, is always in the process of having been. Monumental history, therefore, becomes antiquarian in the preservation of what has already been. Furthermore, given the way the past and the future come together in the moment of vision, a simple today is not presented, for the opening up to possibilities in a future way makes it impossible to conveniently demarcate the today or present from the past and future. Engaged to the future through resoluteness, the present cannot be observed from the position of a spectator. The character of today is negated (*Entgegenwärtigung des Heute*) and a critique of the present is established.[69] Historicity not only makes history (as a science) possible, but the unity and continuity it manifests through resoluteness and repetition also becomes the ground for unifying the science of history.[70] The ontological source of history (as a science) is then demonstrated by the themes delineated by Nietzsche. The primary consequence in this thematization of history, according to Heidegger, is the formation of the hermeneutical situation to which *Dasein* is open in resolutely disclosing its being in the process of having-been. The nature of truth in historiography takes its lead from the historical nature of *Dasein*. Consequently, what concepts are employed in this science and what objects are studied are determined by existential structures that derive from the historicity of *Dasein*. [71]

Dilthey aspired to the explication of historical reason, but it was to Count Yorck, an associate of Dilthey, that Heidegger looked for the crucial difference between the ontical and the historical, which entails an ontological-existential foundation for the *Geisteswissenschaften*. In the correspondence between Yorck and Dilthey, Heidegger recognized an understanding of Dilthey's researche, on the part of Yorck, of which Dilthey himself was not fully cognizant. Through his clarification of Dilthey's ideas, for example his idea of types, Yorck was able to distinguish grounds for the construction of a logic to govern the body of disciplines embraced by the *Geisteswissenschaften*. Yorck was able to see that the historical sciences were dominated by discernment of patterns along corporeal lines, while Dilthey's concept of history was predicated on the idea of a "nexus of force." The concept of pattern is not applicable to such an understanding, according to Yorck, without adequate explanatory bridge principles. Based on the difference between the ontical and the ontological, Yorck considered Ranke and his associates to be mere "ocularists," convinced that what passes in history can never again become actualities. His whole school, on these terms, was not at all historical, but an antiquarian one construing matters aesthetically.[72] On the other hand, all history that is really alive (effective through time) is a critique.[73] The nature of history, therefore, takes its shape from the particular character of *Dasein* and not from the traditional objects of the historical sciences. The central point of historicity is that the primary datum of history is something that lives, not a simple past or abstract ego, but a Self that is historically constructed. The true objects of history can only be living epochs (i.e., *Dasein* as the naturally constituted *epoché*) and events that are not merely destined for preservation but active in their own future and the perpetual source of self-inquiry on the part of the historical thinker. By virtue of the historicity of self-consciousness, then, philosophy that is separated from historicity is inadequate. In Yorck's original words: *Das Selbstverhalten und die Geschichtlichkeit sind wie Atmen und Luftdruck und es mag dies einigen Massen paradox klingen die Nichtvergeschichtlichung des Philosophirens erscheint mir in methodischer Beziehung als ein metaphysischer Rest.*[74]

The historicizing of philosophy has its roots in the historicization of self-consciousness. Because living means to philosophize, according to Yorck, the immediate consequence of philosophy is a philosophy in history.[75] "The separation," therefore, "between systematic philosophy and presenting history is in essence incorrect."[76] Based on Yorck's remarks, Heidegger concluded that a genuine interest in historicity is dependent on working out the generic difference between the ontical and the historical. To conceive the historical categorially is directly connected, for Heidegger, with the basic issues of *Lebensphilosophie.*

## The Historical-Critical Nature of Human Existence

Heidegger's ontological interpretation of history is a theoretical complement to Ernst Troeltsch's logic of history. Troeltsch, in elucidating the manner in which historical inquiry is always the result of the 'own situation' of the observer, has clarified the essential critical function of history. In his own way, Troeltsch has argued that the historical roots of self-consciousness must be the stepping stone for historical investigation. The moment of originality and the function of spontaneous reason (*momentane Vernunft*) remain the pivotal focus of history. In light of the weight of tradition, in the form of effective-history, history is always alive and never merely artifactual. Furthermore, the relation between the formation of the future and the moment of originality in Troeltsch's philosophy of history is analogous to the way in which the future outlook of *Dasein* makes the present a "moment of vision" in which it is denied the status of simply-being-alongside or present-at-hand. The transcendental function of the *Augenblick*, in Troeltsch and Heidegger, flirts with the eternal in its assumption of the three facets of time, but, more concretely, it elevates the historical thinker to the position of a critic. In

Troeltsch this clearly signifies the lively connection between criticism and creativity. According to Heidegger's existential analytic of historicity, self-appropriation becomes the form of the creative which originates in the critical moment.

Troeltsch has defined his logic of history as creative, especially as it continually aims at the formation of the incalculably new. The logic of the new and creative embodies both the emergence inherent to historical dialectic, and the implicit critique of rational metaphysics. Heidegger makes no effort to link the essential critique at the core of historicity to the creative, but *Dasein*, as the clearing for Being, equally challenges the ability of rational thought to grasp the historical world. The authority of rational thought must give way to a more ingenious way of thinking, if the problems of emergence, the incalculably new, and the irrational are to receive philosophical treatment.

The essential critique at the heart of the historical, for Troeltsch and Heidegger, unites them in deriving the science of history from the explicit historical nature of human existence. The temporal unity of the historical nature of human existence makes it impossible to do philosophy without doing philosophy in history. In effect, the historicity of human existence has become a category for Troeltsch and Heidegger, upon which philosophy must turn if it is to be true to life.

Troeltsch's ambition to effect a present synthesis of culture is not met by an equal attempt at prescribing historical outcome on the part of Heidegger. He remains true to his bold ontological approach and avoids all normative pretense. However, in his analysis of death and the resolute appropriation of possibilities for *Dasein*, we are left with neither a subjectivism nor a nihilism but a scheme for individuation akin to Troeltsch's proposal of historical totalization. The unifying structure of both Heidegger's idea of the whole *qua Dasein* and Troeltsch's historical totality, as the meaningful historical unity, stem from the essential critical function of historicity, which necessarily precludes relativism. The historical nature of human existence is conditioned by an evaluative structure, and joins the heterogeneous moments of existence into a continuity that frees historical happening from naturalistic explanation and determinism. Seen as either self-appropriation or self-creation, the historical-critical structure of human existence is a productive, interpretative movement centered in the moment of vision.

From Vico's historical application of the ancient concept of knowledge *per caussus* to Heidegger's proposal of self-appropriation through repetition, we can trace the persistent claim that the world most conducive to human understanding, that is, the historical world, is one that is made and not simply discovered. With this twist, the object of history has shifted from the simple past to the self-decoding spirit of humanity. Moreover, the fact that this world is made and not merely discovered argues for the validity of a more plastic idiom than rational discursive argument to capture its essential truth.

Troeltsch and Heidegger do not lead us into a morass of subjectivism but introduce us to the creativity of the "moment of vision" wherein a unified totality is always presented for critical interpretation. Especially as human existence is comported toward its finality, we confront, without a doubt, the resoluteness of taking action.[77] Heidegger's analytic brings us face to face with our individualized potentiality-for-being. The historicized consciousness of humanity is at the same time a process of individualization. In Troeltsch's scheme of things it is regarded as a duty, but with Heidegger "there is erected a joy along with this possibility."[78] To paraphrase T.S. Eliot's remarks in *Ash Wednesday*, there is rejoicing in having to construct something upon which to rejoice. This is a certain romantic attitude that in many respects initiated the historicist turn found in Troeltsch and Heidegger.

Romanticism must take credit for the joyful response to finitude and the ubiquitous abyss of obscurity at every turn. And so we find in Wordsworth's *The Prelude* how the abyss can stir one to action:

> Hitherto I had stood
> In my own mind and remote from social life
> (At least from what we commonly so name)
> Like a lone shepherd on a promontory
> Who lacking occupation looks forth
> Into the boundless sea, and rather makes
> Than finds what he beholds.
> Lines 510-516, Book III

To gaze upon Caspar David Friedrich's *Der Wanderer über der Nebelmeer* (1818, Kunsthalle, Hamburg which has found its way onto many a book cover in recent years) is to see the perfect illustration of this attitude that equally suggests a correspondence between God, the Creator, and *homo faber*. The abyss that Friedrich's traveler faces, after reaching a summit where he is surrounded by snow-capped peaks, is not disconcerting but an object of contemplation and an opportunity for creation.

Sartre also confirmed the correspondence between God and humankind as an appropriate theme of existentialism.[79] Along with the established correspondence between Heidegger's existential-ontological analytic of human existence and Troeltsch's historicism, then, it is not altogether surprising to equate existentialism to historicism as defined in the predominantly German tradition.

This legacy is perpetuated in Hans-Georg Gadamer's work, which represents a clear appropriation of Heidegger's analysis of human existence. Gadamer's primary historical conceptualizations reflect a genuine concern for the historical character of the hermeneutic situation, as outlined in Heidegger's *Being and Time*. Since Gadamer was deeply concerned with articulating the historical evolution of the temporal elucidation of consciousness, there is some compatibility with Hegel's understanding of the dialectic of history. This rapport becomes especially obvious in a consideration of Hegel's delineation of the ways of writing history within the context of Gadamer's concept of *Wirkungsgeschichte,* which, as we have seen, enjoyed a long history before Gadamer's analysis in *Truth and Method*. In making this turn I think we will find a place for Hegel in our understanding of how it makes more sense to talk of *philosophy in history*, rather than to speak of *philosophy of history* with its prophetic voice. Gadamer, however, will want to deflect the synthetic power of Hegel's logic to make room for a "reflexive-self-grounding," already conceived by Heidegger. The connection between Hegel and Gadamer will test Gadamer's assertion that historical writing would be empty if it did not address the study of effective-history. In Gadamer's understanding of effective-history, one can see a further articulation of the temporal complexity of human existence and how the problem of a universal hermeneutics is derived from this complexity.

# 4

# Mixing Philosophy and History: The Hegelian Roots of Historical Philosophy

## Vestiges of Hegel's Cosmic History

The theoretical problems associated with coherently mixing history and philosophy are addressed in Hans-Georg Gadamer's concept of effective-historical consciousness (*Wirkungsgeschichte Bewusstsein*). This alliance of history and philosophy in Gadamer's hermeneutics calls for the elevation of history to ontology. The decisive turning point in this transition was Heidegger's interpretation of the hermeneutic circle. Formerly conceived as a methodological issue concerning the relationship of the part and the whole in deciphering texts and arriving at meaning, the hermeneutic circle, because it is produced by us, like tradition, evolved into an ontological facet of understanding. In other words, the activity that is produced by the co-measuring of the part and the whole which is vital to interpretation, was declared to be the kind of being possessed by human existence.

It was Schleiermacher who first proposed the universality of hermeneutics on the basis of the pervasiveness of misunderstanding; but within the framework of Heidegger's and Gadamer's philosophy, its universality stems from its integral role in human existence itself. Hermeneutics, that is, is an activity that human existence is absorbed in, it needn't be initiated by external causes, like an unfamiliar text or a work of art, or a communication of any kind. Not only does Gadamer's hermeneutics outline the details of such a metaphysics of interpretation as derived from Heidegger, but in the process he finds a place for Hegel's interpretation of history, especially Hegel's metaphoric examination of the self-revealing Spirit in time in the *Phenomenology*.

The deep imagistic value of the moving-portrait of the Mind in Hegel's philosophy is explored in Donald Verene's study of the pictorial character of the *Phenomenology.* [1] This particular analysis requires that the *Phenomenology* be regarded as an autonomous work and not a facet of Hegel's other systematic enterprises like the *Logic* and the *Encyclopedia.* This interpretation of Hegel is not a whimsical view, since Hegel, himself, refers to the structure of the *Phenomenology* as a gallery of pictures (*Galerie von Bildern*) in the book's final paragraph.

It was perhaps Herbert Marcuse's reading of Hegel in *Hegels Ontologie und die Grundlegung einer Theorie der Geschichtlichkeit* (1932) that first revealed a side of Hegel that resembles the life philosophy of Dilthey. This interpretation entails an understanding of experience as an inner world of effective historical relations and meaning. Hegel was clearly not preoccupied with the concept of human historicity as Dilthey and Heidegger were, but the *aufgehobenes* structure of his historical logic signifies the reciprocity of becoming and meaning, though it is in the face of a cosmic culmination to history in contrast to the individuated sense of completion in true life philosophy. This cosmic completion, however, is specified as the development of the universal individual by Hegel, or the *Bildung* of self-conscious spirit.[2] This cultivation or pictorial construction (*Bildung)* proceeds in the manner that ontogeny recapitulates phylogeny, and so it entails a process of calling up previous content that has already shaped the substance of the mind. The interplay between the individual and the general, ultimately directed at convergence, is conceived as a form of recollection in which the old shapes and forms of thought are gathered up for the individual consciousness which must pass through the stages of the general mind. What alarmed philosophical historians about Hegel's approach to history was that his method (the interplay between the individual and the general) eclipsed methodology in favor of ontology. This thought, however, has shaped Gadamer's comprehensive hermeneutics, and so he has knowingly resuscitated Hegel's approach to history.

While Heidegger's later philosophy is overt about the role of poetry in respect to the "end of philosophy" or at least conventional metaphysics, and also raises the issue of the reality of the productive imagination, Hegel's philosophy, in contrast, is obsessively layered with talk of the self-revealing concept. Between the layers of his rationalistic obsession there is evidence of the historical as the self-completing, self-transcendent reality forever transforming itself upon itself. This dialectic (*aufgehobenes* structure) is integral to the self-transcending reality of Augustine's and certainly Heidegger's temporality. Indeed, Hegel characterizes the manner in which one image supersedes another in the landscape of the mind as a process of re-collection *(Er-Innerung)*, in which the past is systematically internalized and never really lost. This process of re-collection is a necessary step in Hegel's understanding of conceptualized history, which is the final organization of history into a totality. Heidegger also intended his concept of repetition *(Wiederholung)* to refer to the inner movement of temporal consciousness, which is always moving towards a completion of itself. Needless to say, the completion to which temporal consciousness is directed is an individuating process involving personal existence; it does not purport to have the cosmic dimension that Hegel's history possesses. Both movements are an internalization of a projected completion which is generated by an appropriation of what has already transpired, and so both Heidegger and Hegel attempted to clarify the inward historical motion that is forever directed toward totality.

Curiously, death for Hegel represents a powerful force in the understanding of being as it does with Heidegger. It can be safely assumed that Heidegger passed through the preface to the *Phenomenology* and so he may very well have been inspired by Hegel's positive assessment of death as that negative force which the mind endures in order to maintain its being. Death, according to Hegel, is representative of the power that derives from facing the negative, which is not easily dismissed like something false, in an effort to pass onto something true.[3] Remarkably, by looking death in the face and dwelling with it, as Heidegger also recommends, the negative is converted into being. This is also the fascinating move that Hegel makes in his *Logic*, and we can recognize its phenomenological extension in Heidegger's analysis of being-toward-death. The power of the negative is seen as the process of mediation itself which makes something the possession of self-consciousness as with Heidegger's analysis of *Sein zum Tode*.

Donald Verene's analysis of this motion suggests a further parallel to Heidegger's philosophy. When analyzed in respect to the flow of images that passes through its apprehension, consciousness is always faced with a picture before there is conceptual formation. If the picture, *Bild*, comes before the concept, *Begriff,* as Verene argues of Hegel's *Phenomenology*, it must be seen as ingeniously producing the concept in anticipation. If there is only true (absolute) knowledge with speculative or conceptual apprehension, then the images that pass through consciousness are a reflection of this final apprehension. Through what Verene calls the "transcendent optics of the situation," the morphology of images presented through consciousness overcomes its limitations in respect to possession of knowledge by pointing to the unseen.[4] The "cunning of the image" (Verene uses the Vichian concept of *ingenium*) is its ability to produce the speculative apprehension of the whole. Heidegger has worked out a similar "transcendent optics of the situation" in relation to the ability of *Dasein* envision a totality that is yet to be actualized. Heidegger, having been fed by the nineteenth-century preoccupation with delineating nature from history, has reduced the apprehension of totality to that unique historical entity *Dasein*. But, if Hegel's emphasis on the interiorization of the activity he defines as remembrance is like the gathering of an historical movement that is naturally dispersed because of the way that the past, present, and future appear as demarcated from one another, then it is indeed similar to Heidegger's rendering of the process of *Wiederholung*, which also unifies the facets of time that would otherwise remain dispersed.

## The Temporal Horizon of Understanding

Heidegger's existential analytic did not simply promote a methodological option founded on the temporality of *Dasein*, it is the ontological clarification of time as the kind of being of understanding. This temporal foundation, as formulated by Gadamer in *Truth and Method*, calls for a new relationship between subject and object that deviates from both Kant's and Dilthey's aesthetics and epistemology. This new synthesis is an adaptation of Heidegger's ontology of understanding which is first treated in *Truth and Method* within the context of aesthetic being. The explication of aesthetic being is realized in the concept of play which entails a spontaneous movement that always calls for a response. The game plays the players rather than the players play the game, since the position of the player is always absorbed by the movement of the game. Play and equally art as representation, according to Gadamer, is transformed into a fundamentally historical structure. In other words, play or art takes on the form of a totality or whole that only receives its meaning from a particular activation of a game or instance of art. This openness to the other, or the relationship between part and whole in the structure that is only defined as a whole from the vantage point of the particular, means that this kind of reality (of play or art) "stands in a horizon of undecided possibilities."[5] Applied more restrictively to written texts this analysis indicates that understanding is rooted in what is said, which is always something that proceeds from already being concerned with the same subject.

Why Gadamer has singled out the phenomenon of play as illustrative of the hermeneutic situation is not difficult to see. The movement typical of play, most importantly, embodies the to-and-fro movement that is at the bottom of the hermeneutic disciplines. It is also the movement that both Hegel and Heidegger have uncovered at the seat of consciousness; that is, human existence is perennially occupied with the imaginative projection of totality from the vantage point of incompletion which lies within the circle of envisaging the whole. Like the tendency of *Dasein* to repeatedly move toward a goal that would conclude the kind of existence it possesses, play is not simply directed at bringing itself to an end. The to-and-fro motion internal to play means that play itself activates its own activity; it "renews itself in constant repetition." [6] Play, on these terms, is its own foundation and it needn't be initiated by a subject, it is already in existence. This structure does not presuppose a subject; it, in fact, brings the player into its sphere. This independence of the structure of play–that it does not seek to be at an end, though in the face of an end its to-and-fro movement is initiated–must mean that play stands for itself. Gadamer concludes that this autonomy makes play a model for the act of self-representation [7] *Dasein*, being forever directed toward itself, is likewise engaged in a process of self-representation. But, if these structures, of play and *Dasein*, are conceived as self-representing, then their representing obviously includes something beyond themselves, as the act of representing implies. This leads to a further characterization of play and even *Dasein* as we have incorporated it into our analysis. That is, both structures, if they are representing, must refer to something beyond themselves if representation is to fulfill itself. The conclusion is that these structures entail an audience which participates in the fulfillment of representation.

The fulfillment of representation, which entails an audience, finds its original commentator in Aristotle whose *Poetics* outlined the central role of spectators in the presentation of dramas. Aristotle's well-known point of view is integral to Gadamer's analysis of play and suggests that the structure typical of play, *Dasein*, and now drama, receives its completion only when it is open to the spectator.[8] The self-transcending facet of *Dasein* means that the first spectator to receive and complete *Dasein* is *Dasein* itself, whereas in the case of Gadamer's structure of play and Aristotle's aesthetics a separate and distinct subject beyond these structures is required. In any event, the act of representation underlying all these structures implies that they are repeatable. For example, Shakespeare's *Hamlet* has an original text, but it has given itself up to innumerable re-productions. In a manner of speaking, it is never finally complete, though it is forever comported toward completion. The structure that all these realities share, *Dasein*, play, (Aristotle's interpretation of) drama is an historical structure whose permanence (constancy) must be construed as entailing indefinite interpretations which are not really tacked on as supplements to the structure but originally come with the structure itself. From the hermeneutical point of view "texts" must be construed as living experiences that originate as self-constant structures perpetuated by future, undecided possibilities. With the initiation of a "text" (here understood as an event or object that is organized by meaning and purpose) there is always introduced a horizon that by virtue of distance (between the original "text" and a yet to be spectator) development necessarily becomes part of its meaning. *Dasein* embodies this structure and so the distance between its completion and whatever particular content it envisions at any one "moment" also signifies that its meaning is dependent upon development. Like the structure of play which becomes a meaningful whole each time it is played, *Dasein* summons its completion through anticipatory resoluteness in the face of being-toward-an-end, when each time it faces the reality of being thrown toward possibility (by activating a definite possibility). In fulfilling a possibility *Dasein* only confirms that it is primordially comported toward completion or being a meaningful whole. The variety of possibilities that may theoretically complete a "text," as defined in relation to the kind of being of *Dasein*, means that the interpretation internal to self-representation continually re-produces itself so there can be no one correct interpretation from which to derive possibilities.

There is, then, a temporal continuity that joins the process of self-representation, for, though an original creation must always come from the past, it is also accompanied by an ever-present situation from which its originality is filled out. The activity proper to the ever-present situation, that is repeatedly filling out the original creation, signifies that the past is always stretched into the interpretative present. The ever-present situation (the ever-present) makes the past contemporaneous and this is the crux of the theoretical historical point of view to emerge alongside historicism in the nineteenth century. Benedetto Croce's maxim, "All history is contemporary history," reflects the methodological interpretation of this historicist posture. But, more importantly, Gadamer, in disclosing the temporal structure of aesthetic being, that kind of being proper to a work of art (a self-representing entity) has reiterated Heidegger's claim that temporality, as the footing of *Dasein*, is really the preeminent disposition toward understanding. And, not surprisingly, the work of art qua play, because its origin is necessarily attached to its representing function which is indefinitely re-presented, also has repetition as part of its nature. [9] Well, Heidegger has located a similar movement in the special way that *Dasei*n happens historically, as evidenced in Division II of *Being and Time*. The inward movement of *Dasein* through which it re-collects its thrown being by way of its resolute attitude toward the future, or not-yet, is the movement identified as *Wiederholung* (repetition). And, in the very same way that *Dasein's* re-collecting or becoming aware of its thrown state, that is, takes over its past through repetition, does not amount to a simple repeating of a past event that is over and done with, the work of art can not be limited to an original construction discretely situated in a closed-off past. The past, then, is never closed off, because it is its nature to become something different, that is, the future. Gadamer reflects Heidegger's innovative assessment that *Dasein* is not merely world-historical, occurring in the space of history, but is in fact radically historical in the sense that it is the space within which history occurs: "It (the work of art in Gadamer's case and *Dasein* in Heidegger's case) has its being only in becoming and in return." [10] The past, by no means, is co-existent with the present on this analysis, but it is the object of temporal awareness to engage phenomena from the past in a way that makes it present.

Gadamer's argument absorbs aesthetics into hermeneutics. With this move he establishes the comprehensiveness of the hermeneutic situation, if culture and history are the fields within which the objectifications of the human mind are produced. If the work of art is not complete in terms of its meaning, as Gadamer's argument implies, then interpretation or understanding is coeval with the original construction. Not only is it co-existent with the original construction, but it sets in motion a process of interpretation that must be regarded as the process of becoming for the original construction. The object's discrete boundaries may not be altered during this process, but its internal structure which is organized around purpose is open to receive future meaning to be continually integrated into its original form. Meaning, that is, is not self-contained, its function is generative or to be the occasion of repeated interpretations. The analogous situation in regard to positive law, as it has been articulated by Giambattista Vico and others, says it quite simply: the interpretation of the law also creates law. The uniqueness of legal contexts must always guarantee that precedents cannot independently guide judgment; they too must be subject to emendations that reflect the living dynamic between universal application and particular circumstances.

The context for creation is an ever-expanding horizon which nullifies the naive posture that an original context is sacred and must not be contaminated by introducing other perspectives. Since Gadamer proposes that creation makes meaning part of the process of becoming, which moves back onto the thing created, and argues against the position that an original situation can remain unadulterated, he obviously leans toward Hegel as he distances himself from Schleiermacher.

Gadamer begins his wholehearted move toward the Hegelian notion of the simultaneous preservation and overcoming of the past with his analysis of tradition, what he essentially acknowledges as the familiar textures of the past which are sanctioned by every historical relation. Tradition does not simply represent an original impetus to understanding and interpretation that will recede in the process of historical research, or will be abandoned because its meaning becomes obsolete, it is the given historical structure comparable to nature upon which attitudes and behavior are oriented. The condition of nature, however, is, as Hegel saw, alienation, as opposed to recollection of tradition which is the condition of history. Being party to all historical movement, tradition is not antithetical to reason; it is the act of reason itself. [11] There isn't a more obvious affiliation with Hegel in the rest of *Truth and Method*.

If tradition, then, is not something that requires an act of deliberation to be activated, it does not possess an independent objectified status which would make it like a discrete piece of knowledge. Rather, as integral to historical relations, which is the kind of existence human's possess, it is really part of us. As part of historical awareness it cannot be like an object that is independent of consciousness, tradition must be a part of consciousness itself. In the progressive development whose final goal is the correspondence of the concept and the object, Hegel called this structure self-consciousness and regarded it as independent of the consciousness of something. There is a tendency, as consciousness develops in Hegel's *Phenomenology*, for consciousness to forget that it has achieved a sort of independence in which the object is subsumed by consciousness and not autonomous. This tendency is overcome through the form of recollection that throws consciousness back onto its earlier achievements in which self-consciousness has been established through the recognition of the other (the object). In this movement consciousness becomes aware that it is active for reasons other than those immediately confronting it. It has a history, which when recalled establishes a unity with former moments of development. There is no real dispute between tradition and the outcome of historical investigation on the terms of the supportive function of the past. The antagonism between historical research and tradition, as some would have it, argues that the goal of hermeneutics is to arrive at a sound method, while, in fact, this tendency fails to see that interpretation is really a process wherein the subject is vitally connected to the past. This does not restrict productivity, it is the ground of new creation. What amounts to the fusion of the past and the present is responsible for the overcoming of temporal distance, a merger that does not require an empathetic leap into an alien context, as with romantic hermeneutics, but a conditioning that preserves the consciousness of the past. Gadamer seems quite dependent on the Hegelian premise of the self-mediating and self-activating reality of historical movement. On Gadamer's terms this Hegelian premise becomes the acknowledgment of the ontological status of tradition which will support his Heideggerian belief that understanding is not simply a methodological concern, but in fact exhibits an ontological structure.

The repeated conversions that consciousness passes through in the *Phenomenology* also require an act of invention, when forms of consciousness grow old. The act of recollection is part of this move forward, but an anticipatory movement of fore-understanding as expressed in Heidegger's analytic is required to complete the circle.

Hegel intended, as did Heidegger and Gadamer, that this circle possess positive interpretative possibilities, as the movement of the past comes into play with the movement of the interpreter. What rattled the philosophical historians in the nineteenth century was precisely this move in Hegel's work. He raised what an ordinary historian would regard as methodology to the ontological premise that consciousness always has a conception of its completion. Understanding is not enhanced by proper methods, it is the mode of being of human consciousness, and it is synonymous with the structure of the hermeneutic circle. This structure does not guarantee total understanding of any particular text primarily because there is an equal tendency, as Hegel first enunciated, to forget the accomplishments of the past. We must acknowledge the possibility, then, of an autonomous, historically projected "text" that is also tied to the past. This tension, Gadamer has argued, is the communicative space where hermeneutics as praxis really emerges. [12]

## History as the Context of Interpretation

In dismissing the naiveté of romantic hermeneutics and its belief in reproduction, Gadamer, relying on Hegel's format for the conditions of understanding, declares that a "text" as historical projection is defined by the "totality of the objective course of history." [13] Thrown into development, so to speak, means that a "text" must still be "writing" itself. The discrete borders of the original "text" are really permeable and therefore mere reproduction becomes an impossibility, since the "text" must be seen as potentially including more growth. Understanding cannot be restricted to reproduction if a creative germ (future occasions of interpretation) is genetically infused into the "text." The process of understanding (the understanding of consciousness or a "text") is really a productive process that is supported by the ontological, temporal structure of tradition. On these grounds the naive belief that the past can be progressively illuminated by acceptable historical methods must destruct. The ontological premise of tradition as the foundation of understanding means that there can be no independent historical object that can be completely understood in every detail. The recollection explicit in historical being is initiated only after forgetting how far consciousness has come. The historical object includes the acknowledgment of one's own historical forgetfulness which means that there is no historical object at all, but a convergence of historical reality (the past) and the contemporary vantage point upon which understanding is centered. This Hegelian understanding of the temporal structure of consciousness led to Gadamer's insight that implicit in understanding was an efficacy that joins the past with the contemporary vantage point and constitutes an effective-historical relation-the true historical consciousness that he called *Wirkungsgeschichte Bewusstsein.*

This insight was latent in the historical research that preceded Gadamer, but coupled with the convertibility of method and ontology in Hegel, Gadamer has raised the discussion to a new logical order. He is not simply addressing the interest of research which aims at clarifying an object of investigation as the source of truth, he has addressed the need to become aware of methodological consciousness itself which demands that the object of inquiry is always more than the immediate appearance of a phenomenon ("text").[14] This demand is the source of Gadamer's debate with those like Betti who hold to historical objectivism. The historical objectivist has to neglect the natural involvement of historical consciousness and the presuppositions therein in order to argue for the objective goals of its critical methods. This approach assumes the legitimacy of the questions that it asks, and does not ask from where the questions originate. This trust in the honing of proper methods denies historically situated consciousness. The awareness of being historically situated is to be conscious that one is within the hermeneutic situation.[15] Gadamer is asking where the questions that we ask (about historically rooted materials) come from? Unless we can ascertain that the questions are the right ones to ask, we must recognize that the asking of the question is derivative of our own consciousness's involvement with effective historical relations.

The illumination of effective-historical thinking sounds like it approaches the ideal of perfect enlightenment in which subjectivity corresponds to objective reality. We should expect something like this if the hermeneutical situation derives from the historically given. In Hegel's words, the given is all that was formerly subjective and subsequently turned into (historical) substance. We can infer, then, that historical understanding must involve a recapitulation of what Hegel's cultivation of consciousness (*Bildung*) has passed through. But, where Hegel postulates the conceptual fulfillment of history in a final merger of subject and object, when self-consciousness and the object it attends become one, our historical existence, as defined by effective-historical consciousness, means that we can never possess absolute knowledge of ourselves, since we are never really complete. Thus the incompletion of personal subjectivity subverts the attainment of radical ideality.

Gadamer has surely borrowed from Hegel's interpretation of the pre-given in historical consciousness, though he expresses the caveat that final self-knowledge is impossible by virtue of our explicit historicity. Being situated within the process of the cultivation of historical consciousness is metaphorically addressed in Gadamer's concept of the horizon. This picture is designed to reflect the meeting of the historical spectator and a situation from the past; not in an effort to understand the truth of the other but to become familiar with the context that originally made the other intelligible. What is initiated is not really a placing of oneself into the context of another but a merger of contexts which is really an acknowledgment of the effective-historical situation. The Hegelian movement of consciousness, wherein there are no closed horizons, since the past is always assumed into the present, and the future, is adopted by Gadamer to indicate that there is really only one horizon within which we move. There is no need to penetrate alien contexts if historical understanding is constituted of one horizon. Being constituted of one horizon, though, does not solve the problem, it only complicates it, for how is this meta-horizon to be construed? Isn't the totality of which Gadamer and even Hegel speak a *terminus ad quem* which is comprised of the continuity of deliberate moments of understanding and productivity?

Gadamer has reckoned with the opposition between the part and the whole in understanding through the mediation of the effective-historical process. How this mediation relaxes the tension between a concrete, contemporary interpretation and an original "text" does not come about by a recursive move to an original "text" in its pristine ideality; the mediation is construed as a continuity where the seeming opposing poles periodically revive each other. The proving grounds for understanding is not an immutable logical ground but a process of latent meaning realizing itself in repeated interpretative gestures which each time gathers up previous interpretations in an intentional move toward not a final meaning, which presupposes progress, but another moment of practical application which really serves as the proof of an interpretation's success. The applicative facet of hermeneutics is perhaps most clearly demonstrated in the case of practical jurisprudence which I will take up shortly.

Naturally, understanding is always directed and so we should expect involvement with another context which might even challenge our own prejudices. This really means that if there is a horizon that affects the present, it must always be growing. Its growth, however, is generated when it encounters another perspective from the past. This potential challenge to a contemporary world-view is really the way anything like a meta-horizon (meta-narrative) is formed (written). It is not pre-given, nor does it have the quality of a transcendental foundation, it is a creation of a context of meaning which is always forming itself anew. It is not guided by an inherent telos if it is the energy of the prejudicial stance of the contemporary spectator that collects the past into a meaningful totality (a unit of meaning). Can anything like progress in meaning make any sense, then, if the guiding telos of meaning is spontaneously produced by an instantiation or cultivation of a particular prejudice? If progress can be so easily dismissed than to what end is understanding directed? It must prove itself on grounds other than its proximity to a postulated final truth. Since the meaning of a "text," as in the case of law or theological doctrine, arises from their application, the hermeneutic situation might only be fulfilled in the practical operation of derived understanding. These two phenomena illuminate the hermeneutic situation, we might agree with Gadamer, since they highlight the project of the historical consciousness; that is, the overcoming of the distance and alienation that divides the subjective interpreter from the original "text" and its context.[16]

## Heidegger's Dispute with Hegel's Conception of Time

For a present context to pass into a former historical context, there must originally be two discrete contexts. This is only the appearance of things, the process where things go in and out of existence. In the projection of one's own historical horizon the encounter with an opposing point of view is necessary, since the other is required if the tension between the "text" and the present is what always precedes understanding. The awareness of the difference between the present and the past is only temporary, for the past is already joined to the present. The hermeneutic situation spontaneously "recombines what it has distinguished in order, in the unity of the historical horizon that it thus acquires, to become again one with itself." [17] The fusion of horizons, then, really means that there is only one horizon, and doesn't this approach the ideal of perfect enlightenment of which Hegel spoke? The dialectic of positing separate contexts or horizons that are recombined in understanding means that such positing is only a phase in the process of understanding that is obviously required to overcome the forgetfulness that our historical situation is really the surface of the historical cultivation of consciousness writ large (*Bildung*).

Gadamer's articulation of this matter, resulting in his famous notion of operative prejudices originating in our effective connection to the past, is the interactive play between belonging to a tradition and being unfamiliar with an independently projected "text," and establishes the space that is the center of understanding and hermeneutics. The thematization of the hermeneutic question, then, is not constrained by logic, since it is located in the space of communicative praxis where the emergence and construction of the new occurs. Having argued against romantic hermeneutics' emphasis on the reproduction of an original act, Gadamer defined historical understanding as a productive attitude, promoting a more plastic approach to the same original act or text. In this regard the problem of temporal distance in romantic hermeneutics had been solved by Heidegger's explication of the temporal structure of *Dasein* which embodies a radical critique of Hegel's position on time. It is a critique that drives a wedge between the source of history as ontology in human existence and history conceived as ontology in respect to the world Spirit actualizing itself.

Heidegger, in a feigned reluctance to criticize Hegel, acknowledged that Hegel, like Aristotle and Augustine, intuited that time could only be conceived as the extension of the mind. Hegel's consideration of time was indeed a return to the earliest evaluation of time as connected with the spirit, as it was an advance over Kant's intuition of time which remains a subjective entity, since Kant merely aligned it with the "I think." That *Dasein* is the source of the generation of time in Heidegger's analytic is in some ways synonymous with Hegel's understanding that Spirit falls into time. We should expect there to be some compatibility between Heidegger and Hegel regarding the idea of time, if Gadamer, a student of Heidegger, has found a place for Hegel in his hermeneutics. This lineage, however, conceals a fundamental break with the Hegelian worldview.

If spirit (mind) is to be conceived as conceptual and time as an extension, then being in time will represent a combination of the non-sensuous and the sensuous. This is the very point with which Heidegger begins his analysis of Hegel's conception of time in which he appears to protest too much. Having already defined time as the ecstatic-horizon within which *Dasein* grasps its totality, Heidegger found Hegel's conception of time too ordinary. That is, there is a tendency to give primacy to the "now" which invites a quantitative analysis of time similar to that of space. Now, Heidegger has also drawn a picture of time as the stretched-out space within which *Dasein* dwells. Well, Heidegger insists that *Dasein* does not stand in history or time the same way an ordinary world-historical occurrence does. *Dasein* is in fact what makes world-historical occurrences possible. If *Dasein* temporalizes temporality, then time cannot be construed as a sequence of "nows." We should also recall that Heidegger's understanding of "dwelling" is not to be construed as spatial but in the sense of sojourning or passing through.

Hegel does not provide such an analysis of time, and Heidegger explains this through Hegel's dependence on Aristotle's view of time. But Heidegger is also quick to acknowledge Hegel's characterization of time as an "abstraction of consuming" which sounds very much like *Dasein's* ability to transcend the mere today in its anticipatory resoluteness which throws it back on its having-been. The present, on these terms, sounds like an eternal now, which is the way that Hegel identifies the present in the *Encyclopedia*, and Heidegger is willing to admit this.[18] Where does Heidegger's dispute with Hegel really lie? It is the same dispute that Heidegger had with Bergson's conception of time as duration, which on the surface sounds very much like his own view. We would think that Heidegger would agree with Hegel's assessment that time is "intuited becoming" but on closer inspection Hegel's "intuited becoming" (the transition from Being to nothing and vice versa) relies upon the ordinary conception of time as a chain of "nows" wherein any one "now" can ideally be intuited as present-at-hand. The resolution of the "now" into the present-at-hand (ideally) means that time is construed as a series of independent, disconnected points which fill up their own discrete spaces. Conceived as "intuited becoming" means that neither the emergence or passing of time has any priority. But isn't this the formula upon which Gadamer's concept of *Wirkungsgeschichte* relies, and isn't it, then, how the interpenetration of the facets of time in Heidegger is to be understood? Heidegger insists that Hegel's treatment is much too abstract to express the vitality of time revealed in the analysis of *Dasein*. Hegel's restoration of the priority of the "now" as a discrete spatial entity is conveyed, Heidegger would have us believe, in the definition of time as "the negation of negation." According to Heidegger, this characterization unequivocally establishes the identity of time with a sequence of "nows." The "negation of the negation" which is the hallmark of the Hegelian dialectic can also be interpreted as the preservation of the past in the face of a subsequent contradictory moment. On this interpretation the past remains alive in the way it is structured in Heidegger's ecstatic stretching along of *Dasein*; it remains a living force whose effects are registered in successive moments as in Gadamer's concept of *Wirkungsgeschichte*.

Heidegger's dispute with Hegel, then, centers around the "negation of the negation" which is, among other things, Hegel's understanding of Spirit progressing in time. It appears to the mind as an intuition and as such is much like Kant's understanding of time as the intuition that governs the internal orientation of the Self. As such it appears as external to or next to the Self. According to Hegel, as the Spirit actualizes itself through the intuition of time world-history becomes a reality. Time, on these terms, is a tool that is manipulated, and this is the potential character of those things present-at-hand in Heideggerian parlance. But is it manipulated by the Hegelian Spirit if it is the essence of the Spirit to appear in time? This query would not satisfy Heidegger, since Hegel has ultimately identified the essence of Spirit as world-history, and this is to miss the even more fundamental point that time grasps human existence, it is not something that is grasped by human existence.

*Dasein* is not to be reduced to simply being world-historical (occurring in history), it is the source of temporality. Though Hegel qualifies the idea of Spirit filling time in world-history as an abstraction, it remains that he has identified the self-actualizing Spirit with world-history. This move conceals the origin of time. Its roots are left unexplained if it is equated with the essence of the Spirit. Heidegger's dispute with Hegel becomes clear when we recall Heidegger's analytic in which Spirit does not initially fall into time, but in which existence is synonymous with the primordial temporalizing of *Dasein*. Hegel, in Heideggerian language, has indeed fallen "from" time in the authentic mode when he reduced time to the actualization of world-history. Above all, Heidegger's analysis revealed that time no longer contained gaps in need of bridging, but was the enduring foundation upon which existence rests.

True understanding, which is always historical, given the temporal structure of human existence, must always reflect upon its own historicity. For Gadamer, this means that the nature of the historical object is the interplay of historical reality and historical thinking. That is, there is no such thing as an isolated historical object independent of historical understanding. As Ernst Troeltsch and Heinrich Rickert argued earlier: history's object of inquiry is not a self-contained atomistic unit as in the natural sciences, but one that emerges through a process of theoretical-valuation which links the historical object to the subjective historical observer. (There is an implicit appeal to the general and necessary in this theoretical-valuation, since [1] the essential is separated from the inessential, making the historical individual valid for everyone and [2] the historical individual acquires an irreplaceable significance. The effective-historical relation in Gadamer's hermeneutics embodies the purposive structural nexus explicit in the value-relevance of the neo-Kantian interpretation of historical representation.) The link, then, between an original event and a contemporary interpreter is such that both objectivism and empathetic projection are undermined as hermeneutic options.

Historical understanding, which is the essence of all thinking and the fundamental hermeneutic situation, then, does not require the penetration of alien situations. The connectedness and continuity of tradition implied in effective-historical consciousness, which never needs to be activated, means that there is no isolated present situation from which one can assume a fixed position to contemplate a former time. The inseparability between past and present is illuminated by Gadamer's metaphorical use of horizons and the fusion of horizons. Given this picture, historical understanding thrives on the reciprocity of the old and the new. This mutual growth or fusion of horizons, then, means that there is no need to project an historical horizon that is separate from the present. The mutual arising of the past and the present, as well as the anticipation of the future in the form of infinite interpretative possibilities, signifies that the hermeneutic circle is a living unity of the facets of time.

If this process of effective-historical consciousness never needs to be activated, it can never be reduced to a simple methodological concern. It is the foundation upon which understanding plays itself out. Naturally this ontological consideration is an argument against the naive belief that explanation or understanding is only possible on the terms of an isolated object of inquiry and an isolated horizon of the present observer. This challenge to historical objectivism and all expressions of positivism was first expressed by Dilthey in his consideration of understanding as the rediscovery of the "I" in the "Thou."

## Gadamer's Ambiguous Alliance with Hegel

The eradication of the barrier of alienation in the fusion of horizons makes Gadamer's hermeneutics a stepchild to Hegel's system, since it approaches an ideal of perfect enlightenment, what amounts to the mixing of the individual and the universal, the finite and the infinite; in a phrase, the merging of history and truth. This partial allegiance to Hegel is evidenced in Gadamer's challenge to both Kierkegaard and Feuerbach who, according to Gadamer, glossed over Hegel's dialectical demonstration that self-consciousness achieves its own recognition through the recognition of another. The landscape of the mind in the *Phenomenology* proved this over and over again: the case of the master/slave relationship is perhaps the most notorious, the instance of the unhappy consciousness the most enigmatic.

The historical nature of experience was clarified in Hegel's dialectic, since consciousness' awareness of itself provides a platform or horizon from which something can be experienced in the first place. Hegel's dialectical explication of experience is the certainty that the historical movement underlying the appropriation of the past or the old is always confirmed by the intuition of the present and the new. Attainment of truth, then, involves the unity that comes from this dialectical combination– the certainty of the self as self-consciousness and thus immediately set apart from itself. Hegel's assessment of experience, however, is ultimately antagonistic to the hermeneutical consciousness. The culminating merger of object and consciousness in Hegel's system, amounting to the achievement of science, or the absolute certainty that comes with self-knowledge free of anything alien, actually closes off the historical horizon so necessary to the hermeneutic enterprise. To close off the historical horizon is to eliminate the possibility of new experiences. This closure encourages dogmatism, in Gadamer's view, and curtails the openness to experience that is promoted by the historicality of existence—that openness which flows from the unity of the facets of time and enables the hermeneutic posture of consciousness.

This criticism of Hegel is certainly borne of Gadamer's appropriation of Heidegger's analysis of *Dasein. Dasein*'s fulfillment, that is, does not come with the acquisition of final knowledge, but with an openness to continual anticipation of possibilities including death upon which no further possibilities are conceivable. The highest form of knowledge to be ushered in by the mutual assimilation of consciousness and object (Hegel) cannot be a matter of (hermeneutic) experience, since experience depends upon the perception of finitude.[19] Experience, defined on these historical terms, never simply concludes with the understanding of one particular object or another, it is always a reflection upon "what can not be done away with," in Ranke's terms ("*was nicht mehr umzustossen ist*").[20] This insight establishes that what is most necessary for experience is the historical ground of not being able to eliminate the anticipatory outlook of finite beings. The anticipation of personal finitude constitutes the primordial hermeneutic movement, since it is activated by the expectation of totality from the position of historical individuality. A totality or completion that only exists in anticipation, for if it were to be realized the light of experience would certainly be quashed, and the living relationship between object and consciousness would be invalidated.

In Gadamer's terms, it was the ideal of the Enlightenment that most blatantly sacrificed the mutuality of the "I" and the "Thou." The non-prejudicial posture advocated by this ideal was not intent on listening to what had been said.[21] The denial of this reciprocity means that the openness to the otherness of the past is falsely transcended in the exaltation of logocentrism. But, historical consciousness is the hermeneutical experience, since it is open to the claims of tradition and does not seek to suspend them at the cost of observing the internal dictates of a cool rationality. The hermeneutical experience, then, is not a mere methodological consideration if it aspires to the steadiness of the "experienced man." The model of the "experienced man," for Gadamer, is the character implicit in listening to the other while at the same time being prepared to accept things that are conceivably antithetical to one's own position.[22] The many experiences that free the "experienced man" from dogma also liberate him from naive positivism that not only dictates what should be listened to, but also claims that human nature is uniform through time and thus always immediately comprehensible. The hubris in these principles denies the true otherness of the past that presses upon our consciousness, and thus it confuses our naturally historical perception which is representative of the structure of experience in general. If the *aufgehobenes* structure underlying historical movement is understood as negating the past, then the otherness of the past is concealed and the readiness for experience that accompanies all intuition of reality is stifled. To cancel the past is to regard it as only once-happening, thus severing its effects so crucial to the concept of historical efficacy. The readiness for experience does not derive from practice in the sense of repetitive exercise in anticipation of putting knowledge to work, it is implied in the connection we have to historical reality. In this context Gadamer has called for a renewal of Aristotle's practical science in which he has identified the difference between *techne* and *phronesis.*[23] *Phronesis* governed the practical judgments intrinsic to civil discourse which were codified in the ancient art of rhetoric.

In Aristotle's *Nicomachean Ethics*, practical wisdom entails a dialogic relationship, as it arises from deliberation of matters that are variable and can be other than what they are, in contrast to science which seeks first principles that are necessary, invariable, and intent on control. Moreover, it is the experienced man, according to Aristotle, who might even lack knowledge of universal principles, who is more capable of achieving ends through calculated and deliberate action.[24]

## Historical Teleology without Completion

Gadamer's ambivalence toward Hegel is typical of Hegel's many critics. This ambivalence arises from the inability to dispense with the teleological coherence of historical movement, while at the same time arguing against the final culmination of history in the mutual assimilation of consciousness and object. The latter condition negates the process of individuation and self-clarification that accompanies all hermeneutic experience, which is predicated on the acceptance of the other, not the refinement of the other. The transcendence of the other is antagonistic to genuine dialogue and only encourages the kind of dogmatism that denies true historical growth. This notion of historical growth, which emerged independently of the concept of evolution, can be discerned first in Droysen's advice to the historian in *Grundriss der Historik* (1858). Here the concerns are chiefly with methodological and practical issues, but they anticipate the ontological turn in Heidegger and Gadamer's work which has raised the historicity of existence to categorial significance.

In perhaps the first historiographical effort to distinguish Nature from History, Droysen foreshadowed Heidegger's and Gadamer's analysis of human existence. He forcefully claimed, like Gadamer and Heidegger, that "Human existence is Mind under the ban of finitude," a condition that makes existence spiritual and sensuous at the same time.[25] This result, presumably, stems from the experience of a lasting present that through the "power of self-vision" we are able to initiate an internal transcendent movement that revokes our finite boundaries. This revocation of our finite boundaries is an awareness; it is not gained through a manipulation. It is the fusing of horizons implied in Gadamer's conception of the *Horizontverschmelzung* and likewise Hegel's metaphysical optics borne of the internalizing vision accompanying the movement of *Aufgehoben*.

More importantly, Droysen's delineation of the contents of Nature and the contents of History is one of the first modern interpretations of the distinction between the natural sciences and the human studies, and clearly derives from Aristotle's metaphysics. In the case of the natural sciences we are aware of uniformity and the kind of unity in which the temporal recedes in significance. There is movement natural to this realm but it is the movement of periodicity. On the other hand we also become aware of phenomena whose movement results in the creation of new forms; the material structure of this realm (the historical) thus becomes ancillary to the emergence of individuality. In a refrain from the effects of Hegelianism, Droysen characterized this continuity as a kind of teleological coherence in which all that is anterior to the present is somehow forged with what comes after it. The grace of this succession is evidenced in the peculiar capacity for this movement to turn in on itself and produce itself simultaneously, an analogue to organic growth that Droysen spotted in Aristotle's work. This internally generated extension deviates from mere repetition of forms in that its continual self-conditioning amounts to a "contribution to itself," *(epidosis eis auto*). (In Aristotle's *De Anima,* Bk.II,Ch.5, 7, this passage distinguishes between two forms of potentiality, that in which a boy can grow up to be a man, which is of the historical kind addressed by Droysen, and the potentiality for a boy to become a member of a profession, a potentiality that requires more than the fulfillment of an inherent *telos*).

This internal connectedness, in which the medium of time promotes a form of continuity advancing on itself, is obviously integral to a genuine conception of experience. Like Gadamer's rendering of the structure of experience, Droysen also acknowledged the readiness for the emergence of the new so typical of historical growth. Droysen would concur with Gadamer's assessment that the hermeneutic enterprise has the structure of a question that conceivably opens up new worlds, since he acknowledged that the endless succession of historical growth meant that whatever is new contained the germ of yet another new that, most importantly, has the assurance of working itself out and ultimately being heard.[26] Clearly, the analysis of the concept and structure of experience flounders without the preparedness to listen to the other.

There is, then, a metaphysical element that runs throughout Droysen's work. His effort to illuminate historical methodology proved early on that historical practice offers insight into the structure of experience. Gadamer having concluded that effective-historical consciousness is the exemplar for the structure of genuine experience is heir to Droysen's seminal work. The negative and positive guidance generated by Hegel is unquestionable, and it is on the subject of the methods of writing history in Hegel's "Lectures on History" that Gadamer might have discovered the model of the playful dialogue that continually ties consciousness together.

## Hegel's Delineation of Historical Methodology and the Ontological Unity of Time

Scholars have been provoked more by Hegel's emphasis on the place of Reason (*Vernunft*) in history, but he, in his own way, outlined the key methodological concerns in writing history. His categories of Original History, Reflective History, and Philosophical History demonstrate how methodology converts into ontology. A cursory look at these approaches will illuminate this theme of the convertibility of method and ontology.

First, original history involves a minimum of transformation in the representation of events, since the historian is reflecting upon immediate events. There is no temporal gap to speak of, and the historian need not trouble over the conflation of consciousness and object; historical figures speak for themselves.[27] There is minimal tension between the spirit of the historian and the spirit of events to be reported. Consequently, the historian is less likely to transform or obscure the words and intentions of historical figures. As Hegel put it, the historian does not speak with a "borrowed consciousness," personalities express their own aims and actions.[28] Events are less likely to be obscured, since the spirit and culture of the observed and the observer lie on the same continuum. What enables the prospect of original history is the mutually arising consciousness of the observer and the observed. The aspect of working out history that is embodied in effective-history equally entails the contiguous relationship of historian and historical event found in Hegel's original history. The contiguous borders of historical consciousness in original history, which expresses itself methodologically, is more apparent than the adjacent forms of consciousness in effective-history. Nonetheless, the conversational framework of original history is the methodological counterpart of the dialogical configuration that Gadamer locates in the ontological structure of understanding.

Hegel's second approach to history, reflective history, contains numerous aspects that more directly bear on Gadamer's discussion of effective-historical consciousness, for movement away from the present is involved, unlike original history which is non-transcendent. Of the four forms of reflective history Hegel's notion of pragmatic history most closely resembles the structure of Gadamer's effective-historical consciousness. Universal history, another form of reflective history, concerns the totality of a country's history and is written in such a way that the material that is surveyed is manipulated by the individual approach of the historian. The result is that the spirit of former times is obscured and significant events are reduced to summary statements. Pragmatic history, however, does not obscure the reality of the past. Indeed, temporal gaps are bridged when the mind, as Hegel puts it, is opened up to a reality by virtue of the voice which comes from the spirit animating the past. This vital connection between the past and the present has the interesting result of divesting a historically remote event of its pastness and making it present. In a manner of speaking, pragmatic reflections discover what is still alive from the past, and so history becomes focused on the present. An obvious consideration, here, are the moral lessons to be garnered from history, but, more importantly, pragmatic history has the structure of a conversation in which the present listener is open to the "Thou" or the other. If the other, meaning the past, is heard in the present, for all practical purposes the past is denied its quality of having-been and all history, on the terms of pragmatic reflection, becomes contemporary history.

Critical history, another facet of reflective history, constitutes the evaluative posture that historians use to examine the validity of historical accounts. It is not history per se, but the history of historical writing. The author dedicated to historiographical interests does not report on events, but on the recording of events. The subjective response to historical narration is prominent in critical history and the factual basis of historical outcome recedes in significance. The final form of reflective history, fragmentary history, is a propaedeutic to Hegel's primary concern, philosophical history. It is a conceptual approach which aspires to the abstraction of universal history, but its scope is restricted to specific cultural fields, like art history, law and religion. In being concerned with the totality of such cultural phenomena, it seeks to reveal internal relations which make such phenomena coherent. The discovery of inherent guiding principles referring to the peculiarity of a people, for example, promotes a transition to philosophical history, since philosophical history addresses the guiding purpose of world history. The contradictions in this philosophical approach to history are renowned and will not be addressed here. It should be pointed out, however, that the totality that philosophical history addresses assumes an intimacy that is reminiscent of original history wherein there is no potentially obfuscating gap between events and historical reflection upon them. Philosophical history, if it were to take on the structure of original history, would not minimize the reality of historical concretions. The design of philosophical history, though, tends to impose the shape of ideas onto historical data and thus historical circumstances are eclipsed by conceptual frameworks.

Philosophical history abandons the structure of experience explicit in effective-historical consciousness, since the other is eliminated in the self-fulfilling movement of the Idea. As both the content and form of history Reason is not only founded on itself, it is its own ultimate purpose. There is no space for the hermeneutic impulse of the existential moment and the voice of the other (the past) to establish the historical dynamic intrinsic to Gadamer's interpretation of experience. The *terminus ad quem* is identical to the *terminus ad quo*, and so the world dissolves into a fait accompli; devoid of the complexity of intentionality.[29]

## Historiography and the Hegelian Critique

Neo-Kantianism can take much credit for the elevation of history to a science, but this project had been viable since Droysen whose influence can be easily discerned in Rickert's seminal work on the elucidation of concept formation in the historical sciences. Heidegger, early in his career, was stimulated by the logic and conceptual framework for history proposed by Rickert. In his earliest rumination on the historical world, which would reach ontological proportions by the writing of *Being and Time*, Heidegger was preoccupied with the methodological framework for history. In his qualifying lecture to the Philosophy Faculty at Freiburg in 1915, "Der Zeitbegriff in der Geschichtswissenschaften," Heidegger more finely tuned the distinction between the natural sciences and the historical or human studies through an analysis of their respective concepts of time. In his clarification of the structure of time in the historical sciences Heidegger adopted Rickert's methodological innovations in which the object of the goal of the historical sciences would disclose the structure of time intrinsic to history. The object of history had been identified by Dilthey and Rickert as human existence as a culture-creating entity. The objectifications of mind or spirit, then, were the objects of historical inquiry, not in the range of all human events, however, for this would make historical inquiry impossible or conceivable only from the vantage point of eternity. The perception and consequent representation of the process of objectification of the spirit through time would only be manageable if the historian's selection from this stream of becoming was organized by an interest which the present held on any effect from the past. This theoretical position had earlier been enunciated by Eduard Meyer who declared that the historian could only be interested in "historical efficacy," suggesting that history is an event of the past whose effects are not limited to its appearance but continues to effect succeeding periods. This understanding of historical causality entails the methodological canon proposed by Rickert who claimed that historical perception was always tied to theoretical-valuation, since the generative effects of a past event produce valuative processes that initiate the conceptual formulation of the historical. For Heidegger, this meant that historical meaning was dependent on the "qualitative otherness" of the past pressing into the consciousness of the present.[30] Meinecke and Troeltsch also formulated this conceptualization in terms of causality which does not imply value judgments but theoretical-valuation that finally organizes history into identifiable totalities. The historicism generated by this interpretation of history did not promote a reckless anarchy, but implied the ontological unity of time in historical conceptualization, since history could only be conceived by keeping the past alive and extending the present into the future. Thus conceived, history is the living unity of the facets of time captured in Gadamer's hermeneutics. The future cannot be excluded from historical conceptualization if history is the memory of things said and done, since active memory would be worthless if not oriented toward application in some imagined future. It is really only the present which must give up its reality in historical conceptualization, for it really takes from the past in order to anticipate the future.

In this early proposal for the methodical treatment of history, Heidegger foreshadowed his later ontological turn in claiming that the historian must overcome time by living through the temporal gap between past and present.[31] The implicit rule of thumb for historical inquiry is an openness to the claims of the past upon our present consciousness. Heidegger thus concluded that the concept of time in the historical sciences does not have the homogeneous character of time in the natural sciences, but reflects the succession of varying historical periods. Thus structured, history becomes a model for explicating the hermeneutic experience which is stimulated by the openness to the other. Philosophy and history, then, are always stimulated by the context of practical communication or the application of knowledge (memory) from the extended past. That philosophy and history arise from being so constituted simply means that their real tasks are to formulate questions from being-in-the-world. This linguistic predisposition, Gadamer has finally argued, leads to the autonomy of language as it emerges from specific life-forms. Reminiscent of Wittgenstein's later philosophy, this consideration evolved from Heidegger's analysis of the origin of the work of art as the fullest of philosophical expressions in opposition to the conceptual impositions of logic. Here the force of poetry, especially, achieves a singular philosophical position, for language is raised to an art; words stand for themselves, they are not subordinate to conceptual fixations. On the surface, this conclusion is antithetical to Hegel's anticipation of what the Concept, in itself, can achieve. However, the hermeneutic accomplishment of the self-presencing of language in poetry, that is, poetry does not use up words as in normal speech or prose but allows words to become and remain words,[32] is the practical realization of Hegel's self-fulfilling unconcealedness of Spirit in time free of the pride of concepts. Heidegger's later linguistic investigations, then, promised a new foundation for thinking that corresponds to the powers with which Hegel endowed the Concept. This connection implies that philosophical thinking entails the wedding of history, philosophy and poetry, for the self-presencing of poetry combines the narrative intimacy of original history and the anticipation of completion embodied in philosophical history. The privileged way that the "word" functions in poetry and history compels us to say, along with Heidegger, that the "word" gives being rather than it is being.[33]

# 5

# Poetry at the Beginning and End of Philosophy

*Poetry and Plato*

In Book X of the *Republic,* Plato crowns his philosophical query into the nature of the state with a grand irony. In declaring artists and poets *persona non grata*, he unwittingly undermined the political architecture he so cunningly presented through the use of similes, allegory and assorted figures of speech, without which the poet is condemned to silence. Indeed, Book X incorporates a myth (the Myth of Er) without which Plato's argument for the immortality of the soul and consequently his epistemology would unravel. This is the point in Plato's scheme where the art of memory converges with dialectical reasoning and philosophical speculation.

The ancient quarrel between philosophy and poetry, which Plato alludes to in Book X, reduces even the achievement of Homer to a kind of game. There is some room in the *Republic,* however, for songs to the gods and poetry that praises famous men, but the common fare of the poet only conspires to arouse the baser instincts. The leeway Plato guards for the social realism of eulogistic poetry must mean that the poetic roots of culture, even for the staunch rationalist, is integral to the formation of civil society.

Plato's antagonism to figurative and emotional language is demonstrated in his representation of the Sophists, the rhetoricians of his day, as the nemesis of Socrates. Plato's tendentious portrait of the Sophists has obscured the democratic and civic virtue of rhetoric, which like Plato's work, relied upon the cultivation of memory and promoted a pedagogical foundation embracing *logos*, *pathos* and *ethos*, and the *ars topos*, the art of mastering the commonplace as a primary medium for argumentation. Though Plato's form of recollection designated rational thought as the medium of truth, there is implicit in his theory of *anamnesis* a restoration of place that is similar to the rhetorical device of invoking the topical. By the time of Descartes, rhetoric was denounced as a whimsical manipulation of language with no claims to truth. Truth was stipulated as the axiomatic ordering of thought. John Locke concurred with Descartes's appraisal of truth, and went even further in denouncing the figurative use of words as provocateurs of the passions, misleading and constituting errors of judgment.[1] Descartes and Locke were in line with Plato's criticism of poets and artists in Book X of the *Republic*, but their insistence that truth is constituted of unadulterated thought overlooks the epistemological grounding in the recollective (historical) powers of human existence. The agreement between ideas that serves as the basis for truth in Locke's epistemology retains nothing of the art of memory that Plato's epistemology embraces. Plato's theory of memory, though, is focused on *a priori* foundations and does not possess the applicative incentives directed toward the future, and confirmed in the present, as in the more practical discipline of rhetoric. Rhetoric's close ties with the art of the historian, that is, the production of and articulation of believable and persuasive accounts of events is tantamount to the demonstration implicit in the discursive idiom of the philosopher, though the standards for truth will obviously vary.

Protagoras's human-measure tenet epitomizes rhetoric's sanction of the applicative aspect of the art of memory, since truth, that which is and that which is not, is determined within the compass of the oratorical framework of inventiveness. Rhetoric's appeal to truth is not established on a correspondence between ideas, it seeks to create a correspondence between idea and reality (historical or otherwise). The place that Plato retains for the art of memory in his epistemology insinuates history into the epistemological process. But the dynamic of memory in Plato embodies a historical presence that imagines the world as entirely pre-figured, while the dialectic and play of memory in rhetoric is not restricted to the past. Rhetoric confirms a present oratorical context and envisions practical future realization.

Further evidence for Plato's appreciation for the mytho-poetic worldview can be recognized in Socrates' s repeated use of the crafts analogy. Above all, the crafts analogy entails the standard of demonstration embodied in the style of narration upheld by rhetoric. Once the craftsperson has commanded a craft, it can then be replicated by passing the knowledge to an apprentice. Moreover, the crafts analogy signifies the epistemological benefits of indirect reference predicated on comparison to activities that grow out of the sensuous acquaintance a crafts person establishes with their craft. The service that this reference to the sensuous familiarity that the crafts person has embodies the Socratic thesis "virtue is knowledge." Through the crafts analogy, Socrates infers that human existence has a functional excellence. This end can be determined in the way that a craft, which embodies a body of knowledge, is necessarily directed toward a completion that mutually arises with the practice of the craft.

The redemptive and foundational role of poetry was likewise important to those historically minded philosophers who delineated the culture-creating powers of poetic discourse, as we have seen in Vico and Heidegger. The disembodiment of rationalist philosophy restricts philosophical discourse, since the sensuous and the corporeal are confined to the obscurity of an obsolete mythological worldview. The hierarchy established in conventional philosophical discourse, however, has not eliminated the sensuous from human interaction. The alienation of modern-day scholastic discourse arises from the eradication of the historical and poetic. Modern philosophy's reverence for technical language, as witnessed in Heidegger's poetic interlude, is dedicated to framing (*Gestell)* for the purpose of manipulating the environment. The practice of framing demarcates entities from one another and thus obscures the integrity and totality of existence. Consequently, a vocabulary which could illuminate Being, the realization that life comes to us in one piece, is covered up. The more poignant aspect of language, embodied in the expression *Dichten*, allows the world to stand for itself, absent the drive to control and manipulate. When Heidegger intended the pun on thick to describe this form of language, he proposed that forms of linguistic expression, besides conventional poesy, could also achieve a kind of saying that extends beyond the mere framing of something that is signified. Heidegger tried to flesh out the way that language is existentially suited to probe the depths of consciousness. In his own effort at verse in *The Thinker as Poet*, he addressed the proximity of poetic thinking to Being. As he put it, "poetry that thinks is in Truth the topology (*Ortskunde*) of Being. The word *Ortskunde* indicates the familiarity that poetry has with the place in which Being resides.

While it may seem like Heidegger was delving into aesthetics in his turn to poetry, he was still occupied with *Dasein*'s proximity to Being.

Dante's analysis of the introduction of Latin as a technical language, that idiom designed to overcome the linguistic pluralism necessary for the construction of the Tower of Babel, presents an analogous situation to the sterility of an authoritative philosophical idiom. Dante, in his fervor for the vernacular, recognized the positive value of a language that grows out of a particular activity. Though the carpenter might wield a different language than the mason, they equally reflect and respond to the sensuous acquaintance each has with their respective responsibility; it does not subvert production or practical application, though it does undermine the possibility of a doctrinaire voice. An overarching technical language, like Latin, not only subverts pluralism and the mediation of the sensuous, it may very well become the exclusive language of a priestly caste, which is removed from the world of practical application. As an exponent of the ancient humanist tradition, Dante, like Vico after him, appealed to the benefits of practical knowledge. This tradition is associated with the epistemological autonomy of rhetoric and the appeal to the topical in which an imagistic and corporeal element activates conceptualization.

When philosophy embraces the historical, the foundational role of sensuous creation is also acknowledged. Heidegger's early philosophy, with *Dasein* as its centerpiece, simultaneously embraces the grand aspirations of philosophical ontology, since *Dasein* is that ontic entity endowed with ontological privilege, and equally acknowledges the explicit historicality of philosophy, since *Dasein* is the paradigmatic historical entity. That entity which is forever comported toward completion in a process that involves that self-transcending dynamic through which it is capable of creating out of that which creates it. How the poetic emerges as philosophical discourse in Heidegger's later philosophy must recognize that *Dasein* is simultaneously posited as ontological and historical.

## The Categorial Significance of the Historical Point of View

The preceding chapters have examined an idea of history that has gone full circle from Augustine to Heidegger. Within this circle the internal temporal consciousness of which both these thinkers speak has become assimilated to historical methodology, especially in the rise of nineteenth-century historicist thinking. The main consequence of this fusion, between methodology and worldview, as Heidegger's work attests to, is that historical methodology really resolves into an analysis of the natural temporal footing of human existence, without which history is not possible. This identity of the historical point of view and the categorial structures of human existence ground the critique of historical reason.

The philosophical implications of the categorial significance of the historical point of view recommends that transcendental philosophy, which has always been enticed by *a priori* foundations, must be reversed in order to reflect upon the actuality of the transformations of consciousness, which derive not only from the appropriation from previous stages of awareness, but from the impulse to form the future. The impulse toward the future has been reflected in the pragmatic proving grounds of the hermeneutic situation; that is, that facet of understanding which is only realized through application. The structure of experience that derives from the hermeneutic stance, as outlined in this work, is composed of the circumspective concern that is associated with Heidegger's work in *Being and Time*, and equally establishes the credibility of the historical epistemology central to Vico's interpretation of history, the principle of *verum et factum convertuntur.*

Heidegger's implicit claim that the structure of poetry can redeem modern philosophy, which has become alienated from the gods, or the ground of being in more secular terms, ought to be assimilated to Heidegger's earlier preoccupation with the historical dimensions of *Dasein.* His primary concern in his early philosophy was to raise the issue of the way human existence is historically thrown into the world. This early preoccupation (in *Being and Time*) was not devoid of the poetic and metaphoric interpretation of existence, which was to become explicit in his later appropriation of Hölderlin's assessment that "poetically man dwells," and the philosophy that is grounded in the question of the "origin of the work of art." Heidegger's explication of Hölderlin's phrase, and this was already sketched out in *Being and Time,* meant that the kind of dwelling applicable to human existence is not spatial; nor is it to be construed as the mere construction of spaces that provide shelter. Though Heidegger did not intend to do away with the traditional conception of dwelling as a kind of building, his appropriation of Hölderlin's sentiment was construed as a kind of meta-dwelling, which allowed dwelling to take place in the first place. Furthermore, it is not simply an appeal to poets and poetry to resolve the perplexing language difficulties in academic philosophy, it is another articulation of the fundamentally aesthetic apprehension of reality that was previously mediated through the historical categories found in *Being and Time.* The ontological status of history, then, ought to be seen in the light of Heidegger's later consideration of the ability of poetry to stand for itself and not simply use language.

## Vico's and Heidegger's Ontologized Aesthetic

In addressing Heidegger's later efforts to outline the manner in which "dwelling poetically" promised a new beginning or the end of traditional philosophy, we should first recall Vico's philological theory of human existence in which the primordial linguistic apprehension of reality was conceived in terms of the poetic. This link is justified in that both Vico and Heidegger sensed that the institution of beginnings is marked by a poetic assessment of reality. This form of apperception amounts to the creation of a standard, or a kind of measurement, which is informed by the sacred breaking into the profane.

Both philosophers regarded poetry as an act, in words, which grounds and institutes the civil world. This is explicit in Vico's thesis of poetic universals as the expression of the originating function of poetry. Equally, Heidegger, especially in his later philosophy, equated poetry with language's capacity to bring Being to the fore. Beyond this, the primacy of "poetic dwelling," for Heidegger, offered liberation from the limitations of conventional philosophical meaning, which is predicated on the coherence of saying and signification.

The salutary effects of *poiesis* promised an overturning of the restrictions technology placed upon the art of language. The dominance of technology, which alienates the maker from what is made, in turn distances the speaker from language. When discourse is restricted to systematic labelling and imparting information, rather than functioning as a productive voice, the fundamental aesthetic apperception of human reality is abrogated. If the material world is limited to exploiting things, they will never show themselves in the phenomenological way Husserl intended in his dictum *zu dem Sachen selbst.* Both Vico and Heidegger stressed the instituting capacity of poetic expression; that is, the ability for the word to bring something into existence. This philosophy of language goes beyond the linguistic tenet that there is an intrinsic structural similarity between the word and reality, as Plato argued for in the *Cratylus*, in opposition to the Sophists' constructivist standard. Modern language philosophy, it seems, is also invested in such an isomorphism. The primal and originating character of poetry signifies that language is productive in a revelatory manner, and not merely organized to be in accord with the conditions of reality. Heidegger's poetic assessment that human existence dwells inside language must surely mean that language is not a mere appendage of human existence, which develops after the fact of existence.

In his "Letter on Humanism," (1947) Heidegger reclaimed the original productive powers of language, which had seized up through the dominance of Western logic's grammatical dependence on the subject/object dichotomy. Western logic has advanced the dichotomous relationship between subject and object, in order to sustain a technical language that aspires to science. This aspiration, Heidegger argued, is based on a fundamental misconception of thinking and its medium, language. Philosophy's self-elevation to the status of science glosses over the fact that thinking or contemplation requires nothing outside itself to act. Heidegger clearly had Aristotle's *Nicmachean Ethics* in mind, which argued for the primacy of intellectual virtue over moral virtue, since it required no extrinsic or extraneous agent for its fulfilment: "Thinking acts insofar as it thinks."[2]

Heidegger, whatever shape his speculation took, always questioned the ability of language, construed as a system of signs, to address the root problems of philosophy. His penchant for metaphor, neologisms, and constructed speech patterns in *Being and Time* later became the crux of his philosophical speculation. With this tendency Heidegger invoked the metaphoric consciousness of the *Urmenschen* of which Vico spoke, especially in his repeated evocation of the concept of origins and homeland in Hölderlin's poetry. Heidegger's early and later philosophy, then, is the expression of a unified vision that is summed up in an ontologizing aesthetic in which dwelling is poetically construed. The fact that Heidegger's philosophy loops in this fashion has been recognized by those who have focused on his turn to poetry, as in the case of Gerald Bruns. The subtitle of Bruns's study, *Heidegger's Estrangements: Language, Truth and Poetry*, is an obvious allusion to a seminal work in analytic philosophy, A. J. Ayers' *Language, Truth and Logic.* Ayers's positivistic account of language was delimited by a verifiabilty thesis; grounded in a theory of truth which ruled out the kind of metaphysical truth Heidegger discovered in the *Dichten* quality of language. The topographical settings and temporal environment, which frame Heidegger's pronouncements, is just the kind of romantic musing which analytic philosophy cannot abide. On the other hand, Bruns and others have rightly construed Heidegger's assessment of Hölderlin's originary poetic consciousness, in the sense of opening up the world, as a prelude to the more fundamental insight that poetry (*Dichten*) is not just a matter of style, but a way into what Heidegger called *Gelassenheit.* The mode of non-willing which allows things to simply be, thus adhering to Husserl's phenomenological legacy.

Vico shared Heidegger's penchant for establishing the metaphysical import of topographical settings. In Vico's thought, the fundamental *topos* of the *gigantes* was the forest clearing that made way for the sun's rays and subsequent golden fields of grain. Heidegger's use of the expression *Lichtung* corresponds to this primordial space; and, metaphysically interpreted, it is the clearing within which being as unconcealedness (and concealednes) is grounded. The sensory topics implicit in both philosophical environments calls for a metaphoric mediation that is grounded in the sensuous apperception of self-presencing reality. Heidegger's appeal to the power of poetry to illumine this self-presencing rests not on the vestiges of the primordial surroundings of the *Urmenschen*, but on the philosophical insight that the perceptions (the there implied in *Dasein)* of historical entities is at once self-reflective, and therefore always embodied in any assessment of reality. Vico's corresponding interpretation of the revealing powers of poetic apperception claimed that language arose as a consequence of the sensible effects upon the imagination. The result was the transformation of spirit into body reflected in familiar tropes like metaphor, metonymy and the like. These figures of speech, then, constitute the building blocks of human communication. They are to be distinguished from the atomistic constituents of language associated with the analytic tendency to construe natural language in terms of formal notations, which tend to parcel language into discrete units of semantic truth. This latter interpretation of (figurative) language sidesteps the issue that language is a *Gestalt* response to lived experience. Metaphors, on the theory that I am now invoking, are not merely constructed of discrete units of semantic truth that when properly analyzed point to discrete units of reality. The fundamental mistake in this conception of language (metaphor) is to ignore the fact that metaphors are holistic, and when dismantled, bear no resemblance to their constituent parts. For example, there are some theories, like the one Monroe Beardsley shepherded, which hold that metaphor consists of one part literal meaning and one part metaphoric meaning. This approach misses the essential point that there is no simple correspondence between a metaphor's parts and the whole to which it refers. Disassembled, metaphors lose their figurative meaning, since they do not begin with figurative units of meaning. The obvious reason is that metaphor is not merely made up of language units but congeals with an emotional and psychological response. There is only one part to metaphor and that is the metaphor itself. Of course, there is the underlying comparative dynamic of two apparently disparate contexts, but the two disparate contexts join to form a totality that is irreversibly established. Once the so-called disparate contexts constituting metaphor are reduced to their original settings, they bear no semantic resemblance to the metaphor, and thus they cannot serve to elucidate the metaphor.

If language is composed of elements that do not admit of reduction to language, then meaning in language must partially reside in what generates language and speech. In the present context the psychic connections joining disparate contexts would assist in abstracting the meaning in metaphor. A further complication arises when we consider the implications of the use of the word "figure" to describe tropes like metaphor. The implication is that these forms of speech assume a vivid corporeal dimension that is not readily reduced to rational principles. To comprehensively explain meaning in language we must go beyond the presupposition that the constituents of language will only mirror elements of reality. The transposition of reality into language cannot simply be restricted to the conditions of reality or meaningful tropes would be beyond the realm of possibility. When the poet asks, at the end of inspired lines, if he is asleep or awake, it is a response to the free-flowing psychic formation of images that do not readily correspond to the elements of reality, though these images might not be any less believable than a naturalist's report on the sighting of a nightingale. A poetic rapture that might be dismissed as hallucinatory is really the reflection of language's tendency to transform and transcend the constituents of reality. Keats's "Ode to a Nightingale" is not a disoriented account of a conversation between a human and a bird, but an example of how reality is always transposed into the self-logic of language, which is in turn tethered to the psychic connections that accompanies speech and language. Though the origin of language has always been a matter of speculation and even a subject to be avoided; it is not a matter of speculation that language is comprised of elements that do not readily resolve into language units.

In an effort to conceptualize Dilthey's delineation of the *Geisteswissenschaften* and the *Naturwissenschaften*, Heinrich Rickert, especially, corrected Dilthey's insight that the human studies should be distinguished from the natural sciences on the basis of the mental entities that are the subject of inquiry in the human studies. The real distinction between the two realms, Rickert argued, was the difference between conceptual validity in the natural sciences and the need to grasp individuating historical circumstances that were always the result of volitional acts and historical circumstances that originate in the value-relevance of a given context. The result was that the human studies do not require a different methodological approach because they pursue an understanding of mental entities, as opposed to an understanding of material bodies, but because history, the paradigmatic human study, is meaningful only on the basis of the values that underlie historical production. The consequence is a methodology that is informed by the value relations of that which is studied in history.[3] The analogue for the understanding of language, which I am provisionally setting forth, is that to systematically study language, or to posit an appropriate conceptual framework, one must appreciate those constituents of language that do not resolve into language units, namely the emotional and psychic forces implicit in language generation and language use. The kairotic impetus, as we might call it, underlying rhetorical speech and any speech act in which emotion can be abstracted, is similar to Rickert's positing of the value-relations at the core of real historical processes. Just as Rickert distinguished between the real historical processes that are tethered to value-relations, and consequently give rise to a form of conceptualization following these value-relations, so too should we appreciate the potentially kairotic (individuating) impetus to language that equally gives rise to the distinction between real linguistic processes and the non-real configurations of meaning that are only subsequently generated. The corresponding result is that a systematic approach to language must overcome the tendency to mathematize language and embrace the value-relatedness or kairotic impulse to speech. The positing of a universal generative grammar, then, is incompatible with the kairotic impulse of linguistic configurations. We need only recall Aristotle's insight about the nature of metaphor: it surely is a form of linguistic expression that is understandable, but this understanding will not necessarily produce metaphor, since it entails a component that cannot be taught. This unteachable element is always the result of an encounter, a lived experience, an occasion that becomes temporally ripe because of an opportune moment.

Both Vico's and Heidegger's assessment of poetic apperception argues for the certitude of concentrated spontaneous speech that arises from experience. In such circumstances, intellectual assumption is secondary. Analysis of such an experience can be subsumed under a rational schema, but it is not equivalent to the certitude of the original sensuous apperception of reality. Logical analysis may offer an explanatory framework for discussing originary experience, but poetic apperception embodies originary experience. This embodiment is not a form of representation, since the experience, it can be said, still resides in the language, as if it were an after-image. This is precisely the facet of poetic speech that makes it autonomous and permits it to stand for itself. It is autonomous from reality since it has assimilated it. Consequently, it possesses a degree of ontological independence; it does not neatly correspond to the reality that instigated it. On the other hand, logical analysis forces the experience out of language. The conceptualization that resides in the relationship a signifier has to something signified does not begin to overcome the alienation implicit in the dichotomy of language and reality; it merely creates another kind of alienation. That the categorical difference between the spirit and sound of language and the concreteness of experience will always remain the foundation of language's relationship to reality can only be relieved with language's own commitment to the sensuous. A scientistic approach to language, since it tends to atomize speech and language, distances language from the *Gestalt* of experience. On the other hand, metaphor's ability to incorporate the part to the whole, as in the case of metonymy, is not removed from the totality that grounds experience. Moreover, metaphor's contingent nature, that it arises from the emotional investment in an experience, imparts a historicist quality.

## The Metaphysics of Grammar

Vico's postulate of a "metaphysics of grammar" sounds especially modern, since it anticipated the twentieth-century preoccupation with the association of linguistic structure and reality. His theory, however, is infused with a genetic quality that is absent in modern language theory. Vico noted that the developmental character of language tends to assimilate new forms. In addition he emphasised the origin of language, which logically follows from the understanding that language is developmental. These two tenets distinguish his "metaphysics of grammar," say, from Donald Davidson's modern explication of a "metaphysics of a logic of grammar." Davidson regards language as a passive entity. This assessment follows from his belief that abstracting the logical framework of language is required in order to connect language units to nature.[4] Davidson's thesis, it seems, assumes that the structure of language is like the structure of reality. Plato relied on this premise in the *Cratylus*. This position neglects the insight that for language and reality to be so aligned calls for an assimilative process that originates in language's ability to assimilate the body to the spirit. That is, the self-logic of language is made contiguous with the self-logic of the experience of reality, if there is a transformational process wherein experience is transposed into language. The most faithful transposition of experience into language is fulfilled when the totality of experience is assumed by the totalizing tendency of language, as in the case of poetic and metaphoric speech. The link, then, between language and reality is established not through the abstraction of like structures, but through the respective totalizing mechanisms of each structure. Language and reality possess explicitly different structures. The mathematicization of language sponsored by analytic philosophy tends to restrict language to imparting information and mere saying. Thus it ignores the productive powers of language that reside in its originating function. The former interpretation of language's creative powers is limited to mere naming, and the latter interpretation of language illuminates hidden relationships, institutes behavior, brings things into awareness, and, in general, functions as the originating medium that maintains and fortifies the civil and historical world. On the one hand, the analytic dissection of language is intent on distinguishing true propositions from false ones; and, on the other hand, language, as the basis for instituting civil structures, is concerned with the emergence of the real. This function of the latter interpretation of language does not require a judgment of truth or falsity but a spontaneous response that sustains that reality.

To construe natural language in an artificial way, when it is reduced to symbolic and/or mathematical forms, is to construe it as passive. We are then compelled to detach language from the fundamental psychic nexus from which it is generated, that is, being in the world. Such a diremption overlooks the generative nature of language, and the ability to spontaneously create new forms on the basis of understanding contexts. Davidson's thesis, following the emergence of linguistic philosophy in the twentieth century, recommends that meaning is assigned to language in the way that referents are connected to signs. Vico's historical interpretation insists that meaning accompanies language, since it is grounded in lived experience. As such this state of sensuous mediation is language in its pre-articulate embodiment. It is not the case that language hangs on this pre-articulate embodiment, it opens on to experience as a seed pod bursts forth into the air. That is, there is a kairotic character to language that forever subverts any attempt to formalize its generative mechanisms with artificial notations. This kairotic element guarantees that language is comported toward the creation of new arrangements that originate in the psychic nexus of being in the world.

Vico's "metaphysics of grammar" begins with the origin of language. This distinguishes his linguistic metaphysics from Davidson's, since Davidson assumes that sharing a language indicates that we share a picture of the world. Therefore, the study of the language will be a reflection upon the general features of reality.[5] Vico did not make such a leap, though he would not dispute the correspondence between language and worldview. Vico begins with the premise that the new science he proposes, like any discipline aspiring to knowledge (*scienza*), must begin with the matter of which it treats. In Vico's case, the object is the ideal-temporal process of humanity, the synthesis of spirit in time in Hegelian terms. With the process of humanity as the object of science, it is plausible that one would probe the earliest stirring of linguistic apprehension, since the origin of such cognitive apprehension will reveal the dynamics of the initial division between reality and thought (language). In addressing the origin of the first language, Vico theoretically began at the point at which human awareness of an external world comes in the form of an idea. In doing so, Vico has isolated that special juncture at which nature and language separate. This point of division must be the primordial link between communication of thought and the reality of being. This concern for the origin of language, then, is a genuine concern for the way language is connected to reality. In focusing on the generative process of language, Vico, appropriately, considers the operation of the modifications of mind.

Vico characterised his brand of linguistic metaphysics as metaphysics of the mind, since it sought to penetrate the moment of inception of the ideal-temporal process. There might very well be a correspondence that is suggested between the earliest spontaneous movement of language and the external world, but it is also the case that the original linguistic apprehension of the world will be subject to further modifications. Nonetheless, to return, as Vico does, to the earliest stirring of structured thought might explain the natural relationship that exists between language and reality. The correspondence will not necessarily guarantee that a study of language will illuminate the structure of the world; what it does promise is that the first awakening of structured thought (the structure that thinking imposes upon language) will tell us something about the way the world is spontaneously synthesized in language. Such an approach, then, is not directed at dissecting natural language into its atomistic parts, or clarifying the grammatical rules which govern the formulation of meaningful propositions. Vico's theory was not an effort to illuminate the logical structure of a natural language, it was concerned with how language naturally occurs.

To sound the depths of language should demonstrate how ideas were first associated with objects or events. The original acts of "saying," that is, what speech, thought and language have in common, brings things into existence insofar as they are differentiated by the ultimate independence of speech, thought and language. It follows that such a linguistic parcelling of reality has as its foundation the emergence of novelty and new creations. Language does not create reality on this theory, nor does it establish an empty isomorphism between language and reality. The world opens up to our kind of existence, which has language as that horizon which confers meaning onto emerging reality. Emerging reality, in this context, is construed as a way of denoting the process of differentiation which is activated by language's capacity to bring reality out of inarticulateness. Davidson's proposed association of language with the world suggests a correspondence or similitude between the structure of language and the structure of the world. No such proposal is made in Vico's analysis of the origin of language, though he will want to show that a similitude resides in the interactive or praxical space between idea and object. The similitude, however, is not that of sameness between language and object, but an imposition of the form of spirit onto an object to which the senses are directed. The similitude is a spontaneous creation, not the realization of a prior relationship.

Davidson's metaphysical realism, like Plato's view, argues that language can reveal a structure of reality that is antecedent to the creation of language. This move circumvents the rich space between language and reality which is the realm of poetic creation, as Vico would have it. This space, in which communication occurs, has also been aptly labeled the hermeneutic situation, since language really arises in a dialogical relation which entails more than one party.

As an exponent of Martin Buber's thought, Heinrich Ott, in his essay on Heidegger's philosophy of language, "Hermeneutic and Personal Structure," offers an insightful challenge to the scientistic approach to language. Borrowing from the unabashed speculation on the origin of language on the part of Martin Buber and Gerhard Ebeling, both of whom proposed that language originated in a dialogical relation and only later assumed a monological form, Ott calls the place in which language originates and resides hermeneutic space. The use of the name hermeneutic space is Ott's more contemporary way of agreeing with Buber, Ebeling, and Heidegger on the point that language originates in an active responding, as in the case of the fundamental linguistic facet of the *Da* implicit in Heidegger's *Da-sein*. Language assumes an enduring presence in the assumption of a thou, following Buber, and therefore can only be systematically understood if the vital relations and effects upon an assumed other, following Ebeling's theological challenge to linguistic scientism, are taken into account. [6] To group these three thinkers in this way is to further solidify the challenge to a simple reduction of the essence of language to language parts, since it arises not on the basis of words that are simply attached to things in the world, but from the permeable relation that exists between world and thing. If, following Heidegger, the language horizon for *Da-Sein* rests on the interpenetration of world and thing, then understanding language must meet the challenge of understanding the existential dynamic where being in the world and encountering objects mutually arise.

The reciprocity between world and thing does not antecedently result in meaning; meaning is dependent on the praxical space between persons. If communication does not take shape through the benefits of antecedent structures, but on the basis of the horizon constituted of being in the world and the structure of objects, then meaning is historically bound to this existential dimension of *Dasein,* and meaningful connections within this nexus arise spontaneously. Following Heidegger, one might say that language speaks for itself. It can only do so because *Dasein* is constituted of the linguistic horizon in which being in the world and object interface.

Robert Caponigri's example of the proud Spartan in his account of Vico's theory of history demonstrates the spontaneity of this interaction.[7] When confronted with the question as to why his city did not have a wall around it, the Spartan exposed his chest without uttering a word. What this gesture represents is not a reference to a common or shared antecedent reality but a spontaneous movement that can only be interpreted by seeing the similitude between an imaginary wall and the Spartan's chest. It is not a similitude that already resides in the world or language, it is a poetic creation that is made credible in the moment the Spartan answers so directly to the question. This is the essence of the kairotic potentiality that resides in language. There is a link between language (hieroglyphic gesture) and object, but it cannot be understood in the scientific way that descriptive sentences (on Davidson's account) are parcelled into atomistic pieces. The Spartan's metaphoric gesture is the creation of a poetic whole that when disassembled makes no sense. The moment of poetic creation, as in the case of the Spartan, establishes a *Gestalt* of meaning that can only be interpreted within the limits that are formed in the relationship between gesture and object. The praxical space, as we are calling it, is the ground on which meaning is projected. The connection between poetic creation, or the *Gestalt* of meaning, is analogous to the historicist claim that meaning arises from and must be interpreted within an emergent context. Equally, the Spartan's gesture assumes the natural properties of a wall, and in so doing overrides grammatical convention, since a man's chest and a wall are semantically delineated. There is no *a priori* foundation for the Spartan's laconic expression, it is borne of spontaneous creation and a "seeing as" experience that assimilates the spirit to the body.

What must remain intact, if the diremption between the ideal and the real is to be overcome, that is, the gap between word and object, is the structure that underlies the interaction of idea and world. Metaphysical realism, based on the apodictic symmetry of language and reality, has ignored the temporal order in which concrete linguistic apprehensions occur. The transformations that the positive application of language pass through is made possible because of history, which is not something outside the human mind, but a production of the human mind. Human existence does not simply occur in history; by the modifications it generates from within itself it is the form of history, it is time. Enclosing human existence in an inflexible linguistic structure that is determined by set laws ought not be the goal of metaphysical realism. Vico's acknowledgement of the temporal footing of linguistic apprehension recognizes another alternative which bridges the gap between idea and reality. In a sense there are no limits to language if the diremption between idea and reality is closed over by generating a linguistic form that very well could have been otherwise and is, furthermore, certain to be transcended in an interpretation of a subsequent action. What the example of the expressive Spartan demonstrates is the ground of the metaphorical interpretation of reality. It is not an assessment of reality that is dictated by prescribed rules. It can surely be understood what a metaphor is, but this knowledge is not enough to generate metaphor. What is always required is something that only emerges from immediate circumstances, a rule, that is, which is never provided by convention, and which always constitutes an addition to the previously formed human reality. This form of poetic communication is not derivative, it is creative.

Vico does not rule out a final apodictic symmetry between idea and world; his science anticipates a final closure but only through a progressive synthesis of time and idea. He advocated what we might call a metaphysical temporal-idealism, a constitutive tendency of finite human existence toward the infinite. No doubt Davidson's metaphysical realism would be uncomfortable with the Vichian postulate of providence to assure the symmetry of language and reality. Still, this can not be an excuse not to examine the praxical and subjective space between idea and object that is always generated by the form of time, which is human-reality-in-the-making.

## Vico's Critique of the Mathematicization of the World

The ambition that the structure of the world conform to the structure of signs and language that can be manipulated and put into place to establish isomorphism between a sign and an object has a modern exponent in Frege and the lingua-mathematical schools he spawned. But the first instance of the modern priority of subscribing to mathematical knowledge and forms to reveal the transparency of the external world must go back to Descartes. Descartes's initiative begins with the promise of the *moré geometrico* and its systematic manner of achieving orderly truths on the basis of axiomatic evidence. Naturally this procedure must begin with an indisputable self-evident truth that will lay the groundwork for scientific inquiry of any kind, if indeed skepticism is to be avoided. However, his procedure concealed a genuine skepticism that should have been obvious in his holding to the dualism that he equally made famous, the mind/body dichotomy.

The incontrovertible certainty of existence was Descartes's conclusion in the *Cogito.* This reasoning stems from an initial indulgence in the doubting of everything, a cunning tactic in the effort to achieve an irrefutable starting point for systematic reasoning. Descartes would have us believe that systematic doubting ends with the doubting of one's own existence, since how would doubting be otherwise possible. The *Cogito ergo sum*, then, is the original axiom that proves the existence of my own being. But what kind of being is established in this move, and doesn't this formula also suggest that thinking causes existence? Descartes is not clear on these questions, but when we consider the mind/body dualism, it appears that the *Cogito* refers to only one part of this dualism—the mind which does the thinking. Vico's criticism of Descartes's *Cogito* centers on this very point. If we are the kind of being comprised of both mind and body, then the *Cogito* should be quite clear on the proof of our kind of being. The *Cogito*, that is, should be the cause of mind and body. It must be both the mind and the body that must account for our existence, Vico reasoned; for, if we were solely body, we would not think (are there unadulterated material bodies that think, the promise of eliminative materialism notwithstanding?). Furthermore, if we were constituted of just mind, we would only be capable of understanding.[8] Vico's insightful conclusion is that thinking is far from being the cause of mind, it is simply a sign that we have a mind. The clever skeptic, then, will not refute the evidence of signs, but will deny that of causes in reference to the knowledge of existence. More importantly, Descartes has eliminated the consideration of the body in his original axiomatic truth. This shortcoming only produces more ballast for any theory of meaning or truth which follows from this original assertion. If the *Cogito* inspires a metaphysics (of mind) that is prior to all other truths, it will most definitely be one-sided if the body is not incorporated into the original principle. When Vico challenged the dogmatists of his day, he put into play the central principle upon which his new science was to be established. Above all, the principle of the convertibility of truth and fact (what is made) rested on the proposition that knowledge is restricted to causes that produce something. Descartes's first principle did not demonstrate what caused existence in his meditation; he merely acknowledged, according to Vico's principle, that we have minds which think. As long as the mind was uncertain of the causes that produced our kind of being, then knowledge *(scienza)* of existence could not be certain. However, the clever skeptic, along with the person in the street, was not denied access to consciousness (*coscienza*) and would therefore have no problem ascertaining their existence. Vico's challenge to this form of metaphysical dogmatism, then, requires that we delineate between knowledge (which is restricted to causes) and consciousness (which oversees certainty). Certainty, for Vico, was the province of philology, conceived not in a strictly linguistic sense but envisaged as the first science which treats of that which human choice is the author (NS, 138). How human choice, which is always challenged with uncertainty, finds its footing is through the bedrock of the *sensus communis,* which is determined by need and utility (NS, 141). Since need and utility need not be thought, nor demonstrated, they bypass the province of knowledge and become seated in the will to which consciousness *(coscienza)* is directed. The *sensus communis,* then, comes to us in the form of judgment without deliberation (NS, 142). Its cognitive jurisdiction is not the determination of truth and falsity, it confronts reality without reflection on these matters, since it is seated in the will, the fount of choice, which in turn creates that which philology seeks to understand. Vico also employed his central philosophical principle to ward off the skeptics who, while accepting the appearance of things, questioned the possibility of real knowledge of the nature of things. The principle of *verum et factum convertuntur* established that the truths to which we have access must be composed of elements that we ourselves have shaped. Consequently, we are in possession of these truths and elementary conditions in a manner that breaks through the skeptic's position that we can only confirm the appearance of things. Naturally, the domain of this kind of science would be restricted to the world devised by human choice and activity. Vico's postulated lexicon was to be founded on the *sensus communis,* the phylogenic recapitulation that is shared at the level of a people, a nation and even a race, and which can be traced through the various ontogenic embodiments a particular word might have assumed through time. Vico does not systematically treat of this notion in the *New Science,* but it is easy to see what he imagined the force of this etymological principle to be, when he occasionally explores the root of a particular word. For example, in tracing the origin of the Latin language to its roots in the forest (NS, 240), the variety of meanings the word *lex* has passed through since it originated in the word *illex.* The history of a word follows the order of institutions.

Had Vico's criticism of Descartes thrived, theories of meaning and philosophical thinking might have taken a different turn in the West. Vico was alarmed at the founding of science upon a principle that would force an axiomatic ordering of reality, as in mathematics (geometry). This tendency, we can agree, might overlook terrain that did not readily conform to axiomatic tenets, like a straight line is the shortest distance between two points, or any other principle with such self-evident force. In this regard, Vico was especially concerned that the privacy of the *Cogito* would mask the validity of artful methods that receive their force through years of development and communal structuring *(sensus communis).* He clearly had in mind rhetorical methods like topics whose heritage not only extended to classical sources, but equally, he argued, extended into the structure of reality. As a disciplined and orderly manner of communicating, topics provides a medium or connecting term in the process of argument that invokes probabilities in support of certain conclusions. The most relevant aspect of this method, which upholds the extension of thought into the structure of reality, was the development of the category known as the commonplace or the typical.

Vico was arguing for more than the development of an eloquent and persuasive style of argument and oratory, since the connecting term invoked the immediacy of common loci that required no appeal to critical method. They were already in place by virtue of a public art whose origin could be traced to a place, a particular location, a space that was at one time experienced by a body and not just a mind. For example, how many arguments, introductory remarks, statements, lines of thought, etc., bring our attention to that infamous location somewhere between Charybdis and Scylla? To invoke such a location does not depend on having experienced it ourselves; nonetheless, it remains accessible to us as a connecting term that says so much. Don't we all get the picture when this age-old connecting term (place) is invoked? It was received from the geocentric world, it has not lost its force in the heliocentric world, and it still holds its ground alongside the latest theory in subatomic physics. The objectivist line of reasoning which has dominated Western philosophy, for the most part, has neglected to incorporate the body into its methods. This is tantamount to resisting the forces of history, the idea of developmental contexts, common sense, and more public methods that embrace the commonplace, that is, the typical that is always available for public scrutiny. The preoccupation with devising more critically designed methods, I believe, is connected with the desire to avoid psychologism, which is too closely associated with the physical constitution of human existence. This ambition can be placed at the feet of Plato, who was also intent on preserving an intellectual aristocracy.

Vico's advocacy of topics and artful methods honed by public interaction were certainly philosophically revitalized in the *Lebensphilosophie* that begins with Dilthey's articulation of the more fundamental manner of experiencing reality, that is, *Erleben*. The concept of *Erlebnis,* which is embodied in the succession of life philosophies on the Continent, expresses the more poignant manner of experiencing reality which does not hold the object of inquiry or interpretation at a distance. The fundamental premise of *Erlebnis* is that the subject owns an experience, it does not become a reality on the condition that it is reported on; it is a reality by virtue of the subject being coincident with the experienced reality. Novel and innovative interpretations of a reality emerge with the incorporated conditions that makes *Erlebnis* possible, since experience is actualized from within the subject, which is always temporally localized. This form of temporal localization was the kind of premise that Vico sought in his explication of primordial poetic consciousness. It was equally the basis on which Rousseau established the bifurcation between the natural and historical worlds in *Discourse on the Origin and Foundations of Inequality among Men* (1755), in contradistinction to Locke who established this same division on the basis of a logical first principle in his *Second Treatise on Government* (1689). Like Vico, Rousseau postulated a primordial order in which the prototypical human existed, his well-known noble savage. Locke's program, on the other hand, begins with a logically conceived notion of the "state of nature" from which the civil structure falls out. The difference, as I am suggesting here, is a difference that emphasizes the bodily assimilation of reality, in contrast to a logical appropriation of first principles. The latter first step ignores a temporally conditioned situation in which a postulated primordial consciousness is naturally united with itself and its environment. The noble savage, as the prototypical human, assimilates its environment on the basis of open instincts, or judgment without reflection in Vico's sense, which provide the noble savage with malleable conceptual schemata through which reality is assimilated. The natural unity that is expressed in the noble savage's assimilation of reality is tantamount to Vico's postulation of the primordial poetic apprehension of reality in the first stirring of the civil world. The noble savage is possessed of a love of self (*amour de soi*), which directs experience from within (as in the case of *Erlebnis*) and does not, like the vanity (*amour propre*) of the socialized man, distract him with a burdensome, conceptual self-consciousness that estranges the civilized man from the self's natural ability to apprehend the self and its environment from within.

The origin of the civil or historical world, according to Vico and Rousseau, begins with a falling away from a state of nature, but this original diremption is the result of a temporally conceived assertion. The "ascent" (assertion) into the historical world, then, must also involve the body as described by Vico and Rousseau. To begin with a logical first principle is to eliminate the body, and, as in the case of Locke, to overemphasize the role of reason as discursive, literally ordered language, which amounts to an imposition of the ideal of the Enlightenment onto an original "state of nature."

No degree of discursive sophistication has displaced the natural inclination to turn external perceptions into poetic and historicized expression. Discursive, descriptive, linear language and metaphorical apprehension of reality are not competing ways of making sense, they are different ways of making sense. The metaphorical or poetic manner of making sense interests us, as it reflects the mind/body unity with which the likes of Vico and Rousseau begin their analysis of the civil world, that world which is uniquely of our making. There is good reason to believe that the primordial consciousness of which Vico and Rousseau spoke remains intact today; we are not simply endowed with the vestiges of poetic consciousness, shadowy remnants of an archetypical consciousness, the poetic norm remains intact as long as something like metaphor is so widely employed. As I have suggested, if poetic consciousness communicates by way of *topoi* or commonplaces which are associated with certain exigencies and behaviors, then we can easily cite innumerable contemporary examples to match the power of the one previously mentioned, Scylla and Charybdis. We acknowledge when we hear the phrase Scylla and Charybdis a series of events and circumstances intrinsic to the said location, like tumultuous waters, screeching noises, a ship in danger, maybe even a pilot tied to the mast with ears protected from the harrowing, distracting sounds, and so on. In the same way we assent to another set of conditions when we assert the expression "Watergate," "Bay of Pigs," "Another Vietnam." We do not necessarily have to use these expression in a line of argument, but we can imagine them so used in order to convince someone of a certain position. They, in their own special way, conjure an image that entails a series of events. Likewise, they can prescribe certain behavior, but that is not an issue here. It is sufficient to recognize that each phrase or *topos* is associated with a definite context that can be repeatedly applied to varying situations, situations that entail sufficient, similar intrinsic circumstances that justifies their use. In pointing to a location, or even an interaction between characters within a given situation, as in the case of "David and Goliath," an originally experienced perception has been transmuted into an internal expression. Moreover, the assimilated picture, can be asserted repeatedly, that is, reconstituted whenever a line of thought or new situation justifies its use. The invoking of such *topoi* demonstrates, above all, that the fusion of text and context is intrinsic to the metaphysics of interpretation which is mediated by an original sensuous apprehension; one that is potentially perpetuated with the repeated fusion of the hermeneutic subject and the "other" or text.

## The Poetic Norm and the Arbitration of Use

The derivation of an expression and the continued justified application of the expression depend upon an original experience of a restricted domain, and the repetition of the circumstances embodied in the original assimilation of the foundational domain. The foundation is not logical in origin, it is an imprinting that is created through an association between the mutual arising of subject and object, which will continue to correspond to an internal expression in the manner in which it was originally created; what amounts to a repetition of the conditions that linked the originary apperception with the foundational domain. This was Ernst Cassirer's formula for the most original accomplishment of human existence. As humans, he argued, we are endowed with a plastic schematic consciousness capable of structuring reality into mythic forms that could thread culture together in the way that *topoi* are consciously recalled in order to create sense. The proximity of subject and object under the conditions that give rise to a figurative apprehension of reality recall Kant's analysis of the synthetic operation of consciousness. But, if we pursue the origin of this metaphorical and transformative consciousness in a postulated prototypical human existence, as did Vico and Rousseau, we are made aware that consciousness forms a malleable schematic, which is textured in such a way that when impressed onto the conditions of reality it absorbs those conditions in such a way that it becomes that reality in so far as it assimilates the substantial construction of that reality. Consciousness is so inclined and prepared to receive reality because of an original disposition that Rousseau acknowledged as the first motive which induced speaking–an emotional state that synthesizes reality into figurative language. This primordial form of apprehension tended to take reality up in a less conceptualized manner, leaving it closer to its physical embodiment. In this praxical transformation of reality, there is an intimacy between subject and object that challenges a strict grammatical interpretation of language on which technical languages and science rely. The subject does not aim to act upon the object; the configuration of subject and object are joined in a mutual effort to realize the original capacity of language to act independently.

The more original and essential framework for thinking and language, as Heidegger stressed in his "Letter on Humanism," requires that philosophy abandon its scientific pretensions, since thinking intrinsically produces its own field of application. The ontological independence of thinking really frees it from having to act within an applied field. It is really Being that lays claim to thinking and language, and so the scientific authority and clout, which seduces philosophy, also alienates it from its original engagement with Being.[9] Along these lines Dante offered insight into the origin of language that posits a poetic norm at the heart of primordial philological apprehension. This poetic norm, according to Heidegger, was to restore the natural standard that language possesses to gauge its own performance. It is a standard that entails the reciprocity of saying and making which is analogous to the intimate relationship of "letting Being be," which occupies the essence of thinking. Indeed, it is to Dante that Ernesto Grassi turns in his book *Rhetoric as Philosophy* (1980) when he argues that contemporary philosophy has something to gain from examining the ancient tradition of rhetoric, which minimized the rigor of logic in favor of standards and methods more in keeping with the validity of the typical and topical. Grassi's thesis is provocative and is essential to the main consideration here, the relationship between poetry and history. The compatibility of Grassi's thesis and the claim that history has the same structure as poetry stem from the fundamental historicity of poetic discourse. The primary connection between poetry and history, as heretofore defined, consists in the parallel claims each form of discourse and apprehension makes regarding the formation of novel and innovative constructions or structures. The new, that is, follows on the heels of both poetic and historical apprehension. An analysis of Dante's consideration of the poet as orator will shed some light on the poetry/history configuration and will also explain the consequences of the domination of one technical idiom over all other forms of discourse, which in turn eliminates all other forms of expression by virtue of an alliance with a fixed methodological scientism.

This traditional affiliation with scientific language and logic is meant to overcome individuality which is always tied to a restricted space and time in an effort to establish the validity of general rules which will encompass a set of individual occurrences. In addition, the parasitic dependence upon the prestige of scientific and logical discourse is dedicated to proof and the establishment of sufficient reasons and evidence for proof. Therefore, the tradition embracing the primacy of poetic language must be excluded, since it has its source in the historicity of human existence, which emanates from the usefulness of the familiar. Figurative language is content with the lack of evidence for a proof, if it can paint a picture or refer us to a commonplace which entails a set of intrinsic conditions that can be theoretically and physically repeated or reconstituted. Beyond this, figurative language, or innovative language in general, rests on overcoming old linguistic forms that do not speak to the praxical involvement of specific contexts. Dante might have had Horace's *Ars Poetica* in mind when he claimed that language arises within the context of the imposition of need, since Horace had acknowledged the debt we have to the death of old forms of language in our issuance of words that have newly been stamped with the mint-mark of the day.[10] The language of scientism neglects the contingency of temporally mediated contexts, and so philosophy's allegiance to scientific discourse must deny our instinctive access to Being that lays claim to language. Horace recognized that the repute of words, as he put it, was dependent on the arbitration of use and practice "in whose hands lies the right and rule of speech" (lines 71–72, *Ars Poetica*, *quem penes arbitrium est et ius et norma loquendi*).

Grassi refers us to two of Dante's writings, *De vulgari eloquentia* and *Convivio*, for an argument linking the poetic structures of existence with the historicist foundations of human nature.[11] His argument turns on the meaning of the Greek word *poesis* which embodies the productive powers of language. Signifying a mode of production that is driven to alter reality, and arising in reference to a practical task, *poesis* has a praxical or utilitarian function. Like history, which continually alters its own structure, *poesis* is not a goal in itself. Dante's definition of language follows along these lines and it is poignantly related using the theme of the construction of the Tower of Babel.

The methodological fragmentation in philosophical circles forces us to choose from the continued utterances of various schools, each competing for access to truth or maybe even to heaven. Dante's model of the Tower of Babel speaks to this present-day confusion of competing idioms. Dante, like Rousseau and Vico, postulated an onto-historical time, the fabled time of beginnings upon which the first stirring of civilization rest. The form this story takes appears to be universal, since it is predicated on an original falling away from ontological bedrock. On a less theistic note, this mythical fall is adopted by Heidegger in his analysis of the grammatical practice of declension, which he traces to the Greek word *enklesis* which means to incline to the side or away from. The point of Heidegger's analysis, as reported by George Vick in "Heidegger's Linguistic Rehabilitation of Parmenides's Being," is to show that every time a particular mood is expressed using the verb "to be," like "I am," there is the suggestion of a falling away from the verb's root meaning as well as a hidden familiarity with the root, in this case Being. In a rather Platonic move, Heidegger acknowledged that every recognition of a particular (mood, in this case) presupposes an awareness of a unity or fullness that reflects a falling from a universal to a particular (mood).[12] While Heidegger has cast the alienation of human existence from Being in strict linguistic and grammatical terms, Dante's rendering of the original fall is recorded in the history of the use of language. Consequently, his model for this original fall is much more dramatic. The Tower of Babel is responsible for the fragmentation of an original language, since, as the story goes, a variety of language forms developed in relation to specific operations performed in its construction. The ambition of the Tower was to reach heaven, and in the process every job related to the necessity of fulfilling this final goal became the source of a particular idiom (thanks to God's imperious attitude). Depending upon the nature of an operation, say, constructing a staircase, it instinctively produced a terminology to suit this function. The process of designation arose in conjunction with a particular task, and, consequently, the division of labor associated with the construction of the Tower became an expression for the origin of human historicity, since expression is formulated within the context of particular praxical movement.

The disintegration that followed in the wake of this original separation of operations, and their respective idioms, was compensated by the development of another kind of language. Constructed in a purely rational way, in order to overcome the historicity of variety, this constructed language was ahistorical and fixed according to determined generative rules. This universal language, Dante had Latin in mind, was to systematically incorporate the various idioms and oversee the plurality of practical endeavor. It would also facilitate the transmission of ancient knowledge which had been obscured by the preponderance of the various idioms. The invention of language derived from the imposition of need and praxis was too contaminated with the will of the individual and the judgment of usage which determines the norms for speech. Dante, however, decried the impertinence of a fixed language, for it was inauthentic and unable to discern the striving of people who work and live. Inspiring true language, which develops in unison with practice, would fall to the poet, since the poet establishes the meaning of reality in a lively connection with the immediacy of passion. The language of the poet is directed by *Erlebnis*, or the kind of experience the respective laborers on the Tower of Babel participated in, an experience that was enlivened from within, thus giving rise to specific utterances and designations that come alive in practical activity. Here we can see that the separation between word and object, which Vico spoke of in the first form of philological apprehension, is also mediated by the manipulation of the body in the service of specific tasks. The historicist manner of conceiving this link is to recognize that the meaning of this form of language does not arise through argumentation, but through a definite relation with specific contexts that embody conditions that are sanctioned by use and corporeal familiarity.

Vico's exaltation of the process of *ingenium*, the inventive uncovering of similarity where dissimilarity appears to be the case, and the revelatory linguistic creation disclosing something new, where the expected could only be accounted for using rationally deductive methods, was first broached among the Renaissance humanists. Cristoforo Landino (1424-1498) and Leonardo Bruni (1369-1444) were among the earliest thinkers prescribing poetry as an inventive method of knowing. Landino, like others, relied upon Plato's conception of "divine madness" in the *Phaedrus* for his argument in favor of poetry as a form of knowing (recollection). Heidegger's reflection upon our ordinary use of various moods that correspond to grammatical declensions is also dependent on the Platonic identity of knowledge and recollection, likewise revealing the humanist foundation of the epistemological relevance of poetry and rhetoric.

## Reforming the Subject/Object Relation

Like Dante, Dilthey pointed to poetic consciousness as the focal point of the historicist worldview. The connection between the poetic world view and the historicist imagination is especially addressed in Dilthey's "Laws of Imaginative Metamorphosis" through which one is able to see the special connection poetic consciousness has with the concept of historical totality. Dilthey's "laws of imaginative metamorphosis" provide a suitable introduction into his world view, since the poetic and historical world intersect in their operation. Where they intersect we find Dilthey's most profound insight into how the historical world and poetic consciousness generate a process of totalization on the basis of an ever-forming psychic nexus, which, I think, anticipated Heidegger's conception of grasping the totality of *Dasein.*

Dilthey's assertion that poetry is the substratum of all lived experience challenged the epistemological construal of reality and knowledge that had previously dominated Western philosophical thinking.[13] Like Vico before him, Dilthey acknowledged that poetry was, in a manner of speaking, the mother tongue of humanity. In challenging the dominant epistemological construal of reality, his assessment of poetry as lived experience goes beyond any speculation about the nature of the first utterances of a postulated prototypical human. Poetry, as seen in the person of Goethe, Dilthey held, did not involve the simple imitation of nature but embodied a transformative and transcending process that differed from the methods embraced by scientific realism. Poetry's grasp of reality goes beyond the construal of mere antecedent reality, since with figurative methods, like metaphor, it possessed the ability to envision similarity where previously none had been perceived.

What appears to be at the root of Dilthey's challenge to the epistemological construal of reality is an even more acute critique of the subject/object relationship with which philosophy is always concerned. It is one that appears in Heidegger's understanding of the fundamental temporal structure of *Dasein* and surfaces in Alfred North Whitehead's metaphysics, especially in his analysis of the occasion of experience or prehensions. All of these philosophers have argued that the subject/object relationship is fundamental to all experience, but they have all challenged the interpretation that this relationship is one between knower and known, a relationship that generally sets the world apart from the subjective inquirer in a feat of abstraction. All of them agree, in their own way, that the real basis of experience is emotional. Whitehead describes it as "the rise of an affective tone originating from things whose relevance is given."[14] Dilthey's concept of *Erlebnis* denotes an inner, lived experience between subject and object that reverberates with feeling. Heidegger continually acknowledged the emotional predisposition for understanding in *Being and Time* and characterized the essential structure of *Dasein* as Care; implying something other than an epistemological connection to reality. The connection implied in Heidegger's understanding of reality has been previously disclosed as a totalizing relationship that *Dasein* has within itself from its inception. There is a psycho-historical analogue in Dilthey's *Lebensphilosophi*e that is at the heart of the poetic transformation of reality, the "acquired psychic nexus." However, it is of some interest that Whitehead chose the Quaker word, (as he put it), Concern, to typify the fundamental structure of experience.[15] In *Process and Reality* Whitehead referred to this structure as a prehension or the activity of appropriating a datum so as to make it the possession of the subject. This connection to reality deviates from any representational notion of perception, since the process of prehending involves a transformation of an impression to the constitution of the subject. If it is on such a basis that the world is made intelligible to subjects, then the customary subject/object relationship must be reformed to accommodate the special way lived experience is really a unity of perception wherein the subject is forever emotionally prepared to transform external impressions into internal expressions. The use of the word Concern to denote the special way that emotion mediates experience is so close to Heidegger's concept of Care to justify the claim that the two philosophers were listening to one another.

The suggestion that the traditional epistemological construal of the subject/object relationship be overhauled to accommodate the more emotional and creative elements of interpretation is, in a sense, what Gadamer appeals to in his metaphysics of interpretation. In finding a place for the sensuous transformation of reality in interpretation, which derives from the natural historical posture of human existence, we can readily appreciate Whitehead's postulation that Concern forges the relationship between subject an object. Such an interconnection is radically different from the promise of the egoless posture advocated by most strains of positivism.

Alla Bozarth-Campbell*'s An Incarnational Aesthetic of Interpretation*, which is dedicated to these themes in interpretation, clarifies the creative process involved in any act of interpretation. Dedicated primarily to clarifying the role of performance and the audience in the completion of a poem, she revives Aristotle's assessment of tragedy which also requires the audience's response for its totality. Though there is scant attention paid to the idea of history, she does address an aspect of time so central to the creative moment in which interpretation takes place. Borrowing from the theological paradigm that she finds in the opening lines to the Gospel of St. John, where the word becomes flesh, Bozarth-Campbell introduces the kairotic aspect of time to which language is forever comported.

The discussion in Bozarth-Campbell's book is summed up in its title, which is further broken up into what she imagines as the three facets of the continued creative process in interpretation. The three facets of interpretation involve an original act of creation of the word, in a poem, which must await its *kairos* or incarnational aspect when it is taken up by performance and embodied in an interpretative gesture. Finally, the audience that receives the word-made-flesh through interpretation is transformed or moved by the power of the word. Bozarth-Campbell's discussion comes with theological ballast, but her use of the religious metaphor suggests the fusion of horizons so central to the reciprocity of subject and object in effective-history. Similarly, the dialogue involving embodiment and transformation in her sketch of interpretation is comported toward completion only in an act of transforming reality or the emergence of the new and novel, aspects of the philological apprehension embodied in the normative poetic aesthetic outlined thus far.

## Poetically We Dwell

When Heidegger concurred with Hölderlin's assessment that "poetically man dwells," he was reviving a claim that he had made earlier in his "Letter on Humanism." In a rare humorous moment, Heidegger explained that the preponderance of the technical interpretation of Being, along with its technical logical idiom, had only established a standard that did not measure up to the task of thinking. As he put it, it was very much like evaluating the nature of a fish after placing it on dry land.[16] In this regard, he anticipated the liberation of language from the strictures of grammar and logic, and this, above all, meant that language must abandon its subservience to the sciences which promised autonomy, while only concealing its genuine praxical nature. In service of the sciences, language and thinking had forsaken their instinctive productive capability. In its fervor to be like the sciences, thinking had only abandoned its essence which would only be restored by poetic creation. Here, Heidegger equated real thinking and poetic creation, since he intended that the true nature of thinking, as implied in the Greek word *poesis*, was inherently active: Thinking in itself was activity and need not depend on serving the sciences in order to be productive. What then is the proper standard to evaluate the essence of thinking and poetic creation and what does it mean to claim that the liberation of language from grammar and the technical philosophical idiom "into a more original and essential framework is reserved for thought and poetic creation?"[17] The implication, all this being said, is that the poetic apprehension of reality (genuine thinking) maintains its autonomy through its intrinsic productive forces; it only loses its practical aspect when it resorts to sustaining the sciences. This remains a real temptation, since the sciences promise prestige and the rigor of a rational framework. At the risk of appearing unscientific and possibly irrational, thinking, Heidegger argued, engaged the scientific paradigm while turning away from the claims Being has on thinking and language.

If Heidegger's concurrence with Hölderlin's insight that "poetically man dwells" is to be aligned with his earlier claims about the function of language and its vital role in the unconcealedness of Being, then we must accept that his approach to language, like Vico's, goes beyond the analytic preoccupation with demonstrating the structural links between language and reality. Both Vico and Heidegger pursued the philological interpretation of existence in an effort to establish a truly historical understanding of human nature. Hölderlin's words were not merely a characterization of the role of the poet, but an insight into the nature of human existence; our kind of existence is poetically predisposed.

The technical idiom has sanctioned the primacy of a factual and abstract conception of the world, since the methodological bias of science presupposes a soothing "atomistic" correspondence between the world and language, which, of course, remains entrenched as long as technology testifies to gains derived from the manipulation of the environment. On the other hand, the theory of language, which supports the claim that we are primordially seated in a poetic mode, maintains the disparity between language and the world. The inherent romantic disposition of this approach to language aspires to a non-manipulative relationship with the world, but it is predicated upon a transformative relationship with the world as we have already indicated. The disparity between language and reality, on this theory, does not mean that our existence stands over and above the world; it presupposes that we are already inside a world, the pre-articulate domain of language which constitutes the original linguistic disposition of human nature.

Heidegger's philological pursuit of human nature, like Vico's, assumed that the essence of being human was distorted by the bias that human nature is fundamentally rational. Heidegger aimed at correcting this prejudice with his personal play on the term existence. He wrote the term as Ek-sistence to capture the ecstatic temporal dimension of human is *Dasein,* and by the time he wrote his controversial "Letter on Humanism," this rendering of thrown temporal existence was intended to overcome the homelessness of humanity in its estrangement from the destiny of Being, a theme he pursued as early as the writing of *Being and Time.* This temporal foundation, which is the essence of human existence for Heidegger, must negate the conventional rational interpretation of human nature since this seating in time, the temporal ecstatic mode, precludes the traditional subject/object relationship maintained by the scientific idiom. Cast in terms of the homelessness of humanity, the estrangement from Being is perpetuated by the scientific paradigm for humanity; in Heidegger's words, human *Dasein* has become the lord of beings and obscured its role as the shepherd of Being.[18] With such a claim Heidegger is clearly taking the risk of sounding irrational, at least against the background of the success of the scientific idiom. However, such a claim reveals the profound humanistic roots that might be traced to classical sources, a method that is to be expected from a thinker who has explored the latent philosophy in our original linguistic make-up. More importantly, this assessment of *Dasein,* which receives no further explanation in the "Letter on Humanism," meant that human existence was in service of Being. The intimacy implicit in the relationship of *Dasein* to Being, which was disclosed in *Being and Time,* was described in Heidegger's later work as an appropriation by Being in which the essence of human existence was only fulfilled in serving Being. This service to Being assumes a form arising from the philological disposition which does not represent or designate, since Being cannot accommodate a label in the referential way that discrete entities do. For Heidegger, overcoming the tendency to designate and pass along information, then, was the primary focus of going beyond the Western metaphysical tradition.

The ecstatic dwelling that was intended by Heidegger in the use of the expression ek-stasis does not confirm Descartes's s *Cogito* as the preliminary stance of (thinking) human existence. Rather, the literal meaning of the Greek term, "a being put out of its place," illustrates that representational thought alone cannot encompass the linguistic sphere of human existence. The most provocative inference that Heidegger makes regarding the temporal ecstatic horizon of human nature is that it constitutes the necessary predisposition toward Being. This would be an expectation of that essence which is constituted equiprimordially of all the facets of time. We can then make sense of Heidegger's claim that *Dasein* exists alongside Being, in the nearness of Being. Heidegger recommended, in this regard, that system building and complicated conceptualizations, typical of that critical philosophy which followed in the footsteps of

Descartes, be understood as that form of thinking with manipulation as a goal. Thinking, he argued, must retrace its steps and activate that form of thinking which preserves a questioning that experiences.[19] Initiating a questioning that experiences does not engage the world in a manipulative way, but engages it in a manner that presupposes any calculation of reality. This reflects the typical Heideggerian questioning of the philosophic tradition. This challenge to the Western metaphysical tradition had been latent since *Being and Time,* but in his later work, especially *Zur Erörterung der Gelassenheit,* is undeniable. In this essay Heidegger analyses meditative thinking, and the clever airing of views in this dialogue also claims the symbiotic relationship between saying and experiencing.

On the surface we are aware of thinking as utterances that reflect feelings, give orders, tell stories and the like. However, the order of thinking that Heidegger speaks of in *Gelassenheit* presupposes the use of language that is not distinct from experience. This can only be so if experience still resides in language. This thinking is obscured by the kind of calculative thinking that excites the will, the fulfillment of ends, and which Heidegger associated with technology and the modern world. The claim that experience remains in language and is not dispelled as in re-presentational language, since it takes the place of experience, echoes the Vichian proposal that the originary poetic apprehension of reality is mediated by the transformation of spirit into body. The result of this originary linguistic insight into reality was the creation of poetic universals in which reference to the external world (and consequently mere designation) is transcended in a metaphysics of the senses and the imagination which linked the typical (universal) with the corporeal interpretive mechanism of the initial "vulgar" attempt to conceptualize. Vico's premise for the origin of poetry and poetic universals challenges the notion that real knowledge emanates from esoteric wisdom that is restricted to the sphere of "great and rare philosophers" (NS, 384). The link between linguistic expression and the physical ordering of the external world is maintained in the corporeal net the poetic universal casts over the realm of the typical. The link is not maintained through a kind of regularity that is uncovered in the external world; the corporeal conceptualization of the external world is generated from within the modifications of the mind since this sense-grounded ordering faculty is innate and does not derive from the external world. It was only in this way, Vico argued, that one could begin to talk of a metaphysics of mind. That this metaphysics was grounded in history and the sentient apprehension of reality must eliminate the possibility that metaphysics can begin with unadulterated rational first principles.

Heidegger intended that this higher order of thinking engaged what is prior to thought. It does not demand technological intervention into reality, but a composed and steady disposition reflected in the German word *Gelassenheit.* The spiritual implications of this frame of mind are obvious and suggest a detachment from the technological access to existence that releases us from a concern for just the material ordering of reality, to a delivering over to the presence of Being. To be released onto the presence of Being can only mean that the will is out of place, and so thinking abides in the original unity of language and reality that was obscured by the scientific and technical governance of reality. The steadiness onto the presence of Being can never be expressed in words that simply re-present since it is informed by the underlying field of awareness; and so the sentient contact with this original ground keeps language from separating altogether from reality as in the case of sheer representational thought which is dedicated to reproducing reality.

Heidegger, in his later writings, framed his assault on the rational and scientific bias in philosophy in terms of the homelessness of *Dasein.* This analysis was informed by his reading of Hölderlin's poetry that throughout deals with the themes of homeland, coming home and the like. Heidegger was emphatic that these concerns were not to be interpreted along nationalistic lines, though there is the strong temptation to do so because of the linguistic chauvinism expressed in his lectures on the pre-Socratics and the blatant nationalism evidenced in his foray into National Socialism. But, when Heidegger speaks of the uprootedness of human existence in *Gelassenheit,* he clearly places the blame on the calculative planning that reduces existence to automation and colors reality with the quality of "standing-reserve" (*Bestand*), a theme that he addressed in more existential terms in *Being and Time.* The loss of human existence's autochthony (*Bodenständigkeit*) is exile from its original steadfastness between heaven and earth; what Heidegger paints as human existence's original dwelling as an *axis mundi.* It is within this kind of dwelling that Heidegger sought the standard that was to be the true measure of the originary soul of language, not within the sphere of the manipulative analyst who perceives nature as a vast inventory in service of modern technology.

The non-representational thinking and speaking which Heidegger addresses in *Gelassenheit,* requires, as the title of this intriguing dialogue suggests, a certain steadiness onto the metaphysical region that overcomes the opposition between us and an object that appears against an horizon. Intervention of any sort is out of the question, since we might also render the word as "letting-be" which suggests an equanimity and an acceptance that comes in the form of listening. Conventional representational thinking is naturally excluded for it merely reproduces through designation. At the same time representational thinking upholds the conventional interpretation of a transcendental horizon through which objects become available for labelling and consequently deployment and manipulation.

The dialogue along the country way proposes a serenity that would frustrate the manipulative analyst, since the proposal assumes that words can never really represent anything. At best they signify that something dwells within the range of its expressibility. [20] This does not frustrate saying altogether; it encourages the transcendence of naming as labelling and a turn to that kind of naming through which the named and the naming happen together. [21] Naturally, this kind of naming or saying requires a composed listening which allows the namable to be named. The activity of naming, then, is not a matter of appellation which at first holds the name apart from the named. The spontaneity implied in the mutual arising of the named and the name suggests that the namable originally houses the name and the named, which become detached when the metaphysical unity of the namable is responded to by the attentiveness of poetic dwelling and the serenity of *Gelassenheit.* The juncture at which the name and the namable detach occupies the space within which poetic dwelling becomes possible; it is only established in the constancy of steadfast waiting which does not impose linguistic assignments, but allows human nature to return to its origins in what is prior to thought. The obvious nearness of the named and naming further implies the privilege of human existence *vis-à-vis* the bringing forth of the namable into living words.

Heidegger coined the term *Inständigkeit* to characterize the releasement onto the metaphysical region that is the source of truth which would guarantee spontaneity of thinking. The word suggests being capable of a constancy or permanence that would assure the steadfastness (endurance) required to receive the nature of truth. Not piecemeal truths set apart from one another, but a reception of truth in its totality. This is all revealed in a poem that the scholar in the dialogue on the country path introduces to illuminate how human nature is released onto that which is prior to thought or the region that regions. That it is the "thinking heart" that is to be cultivated with a humble patience in order to receive the truth intact (as a whole) is a clear sign that belonging to this metaphysical region is not a matter of unadulterated rational thinking (*Nie ein Wahres allein,/ Die wesende Wahrheit/ Heil zu empfangen./ Für weite Beständis,/ Bestell das denkende Herz/ In die einfache Langmut/ Der einzigen Grosmut/ Edlen Erinnerns.*).[22]

The state of enduring in *Gelassenheit*, then, means that human nature is a part of what is prior to thought. Metaphysically speaking, belonging to this region also means that this region needs human nature. The appropriation of human nature is required in order for the metaphysical region to be what it is.[23] To conceive of this region requires overcoming the tendency to objectify a relation between ourselves and the entity Heidegger referred to as *Gegend,* or *Gegnet* in its old form. As an open expanse, this later formulation of the *Lichtung* gathers everything unto itself and with everything else in a restful lingering. [24] Nothing appears to happen in this region because things no longer stand opposite us. More importantly, nothing stands at all.[25] The state of *Inständigkeit* is a vigilant non-willing that rules out, according to the scholar in the dialogue, ostentatious conceit, arrogance, and combativeness. The truth in its totality is not won with aggressive linguistic assaults in which the technical idiom of the day is deployed. The opposite disposition is required to receive the truth intact; it is a demeanor that collects itself and maintains composed in letting-be. Heidegger called it a composed attitude that is gathered within itself. This disposition entails the non-willing state in which the truth-intact is involuntarily summoned by releasement.

Moving along the country path was not a whimsical metaphor for Heidegger; it resurrected his earlier reference to the enchanting lit forest clearing through which the world is linguistically proffered in a movement of concealedness and unconcealedness. The metaphysical region that he spoke of in *Gelassenheit* was also enchanting territory. Its powers are not disclosed by mapping out the territory with linguistic assignments. Its secrets are revealed in the waiting which transcends representation. At once, this waiting is a releasement from willing (designations) and a releasement onto the metaphysical region that is the opening onto openness. It is, then, not the way (*Weg)* itself, but a movement (*Bewegung*).[26]

What can we expect to accomplish in a state where the will is inoperative? To actualize something requires a basis in volition. Without it how can there be resolve? But this is precisely what we should conclude; there is no possibility of action where there is no differentiation between us and objects in a field. To eradicate resolve in this way seems like Heidegger was abandoning his understanding in *Being and Time* in which *Dasein* is resolutely constituted of Care. There is no contradiction at stake here. Resoluteness in *Being and Time* turned *Dasein* toward authentic potentiality-for-Being in which action is not taken, but in which taking action is already present since authentic resoluteness and the Care structure of *Dasein* entails ontological completeness in which all factical possibilities are already present. *Dasein* is temporally constituted as a totality not after the fact of fulfilling a range of factical possibilities; by virtue of being-toward-an-end it is held together as a totality. The unity that rests in the metaphysical region that we have access to through *Gelassenheit* and *Inständigkeit* is a reiteration of the totality that constituted *Dasein* in *Being and Time.* In both cases, totality, or encountering the truth intact, is not gained through taking action, since action presupposes the manipulation of entities in an objectified field. The premise of an objectified field and an independent cognitive awareness, which conventionalizes the field through linguistic assignments and the consequent manipulation of the field, does not accept the primordial unity of the metaphysical region and releasement onto the region in that state Heidegger calls *Inständigkeit.*

It makes sense, then, that Heidegger conceived of language not as a functional appendage of human existence that could be wielded like a tool, but as a region within which we reside. In respect to dwelling poetically, which is not restricted to poets, being inside language meant that language (*qua* Being) uses us; it calls upon us to linguistically frame the range of expressibility. Here, Heidegger draws our attention to the symbiotic relationship between human existence (*Dasein)* and Being. We come onto Being through the correspondence established in the poetic response which translates the movement onto truth into living words. This means, then, that this poetic response onto truth and Being, which Heidegger characterizes as the essence of human existence, establishes translation (transformation) as the paradigm for any further assessment of human existence. This root paradigm, however, does not entail translation from one language to another but the transposition of the breadth of what is sayable into spontaneous thinking. Following Heidegger, there is no simple correspondence between the domain of what is sayable and naming, as the one domain does not really extend into the other. Naming, that is, does not have jurisdiction over the range of the sayable, it is contiguous and primordially near-at-hand.

## "In Lovely Blueness..."

Heidegger's examination of Hölderlin's poetry is summed up in his assessment of the phrase *dichterisch wohnet der Mensch* in the poem *In Lieblicher Bläue...* The poem is a meditation on humanity's relationship to God. Hölderlin begins his reflection with images of a steeple that projects into the blue heavens. The steeple's engagement of the sky soon turns to the question of humanity's reach for the divinity, and the real issue of the measure of humanity comes to the fore. It was sufficient for Hölderlin to raise the question of the criteria that unite humanity to God for Heidegger to uncover the question of how we dwell. For Hölderlin, the measure or standard that could guide humanity is as obvious as the sky. However Hölderlin envisioned the standard, one thing seems clear: the measure is not to be construed in terms of those things that humanity has merited or gained. What humanity possesses is one thing, but the way in which humanity resides is another. That humanity is replete with things it has acquired is secondary to how humanity dwells. The standard to measure humanity, according to Hölderlin, is captured in the idea that we dwell poetically on this earth. What Hölderlin has done, and what has guided Heidegger's meditation, is to have made the distinction between the essence of human existence and what humanity has at its disposal, the things that it has acquired. The measure that gauges human existence, then, is not found on this earth. *Giebt es auf Erden ein Maass? Es giebt keines.* (Hölderlin, section II of *In Lieblicher Bläue...*) This is precisely what we are to derive from Heidegger's dialogue on *Gelassenheit*, since the question of human nature involves understanding something that is, strictly speaking, not part of our kind of existence; but, something that appropriates us. This puzzle directs the dialogue from its inception and is resolved in the same manner that Hölderlin measures human nature.

Human nature derives its essence from the way in which it resides or is ontologically seated; in the same way the interlocutors in Heidegger's dialogue discover that (meditative) thinking is not determined by thinking but through the metaphysical region that is revealed through a vigilant *Gelassenheit.* How human existence dwells first brings forth its nature; it is not determined by mustering accomplishments. To construe the manner in which human existence is ontologically seated (dwells) is to marshal a criterion that is not used to calculate comparison, what we would expect with an emphasis on merits and accomplishments. Rilke's advice to the poet in *Letters to a Young Poet* are *a propos:* the poet must await the creative hour with deep humility and patience; to concern oneself with tallying achievements is a distraction. The poet must be as patient as one looking onto eternity.[27] What Goethe's Faust in his desire for materially rich experiences cursed above all, patience, is everything to the poet.

In *Erläuterungen zu Hölderlins Dichtung*, Heidegger offers an exegesis of Hölderlin's poetry and a detailed reading of the phrase *dichterisch wohnet der Mensch.* Neither Hölderlin nor Heidegger were clear on how this expression applies to human existence *per se.* What Heidegger does offer in this later writing, however, are clues to how his earlier analysis of *Dasein* in *Being and Time* correspond to his exploration of the poetic ground of human existence. This result is most interesting, since it offers a systematic way to view Heidegger's philosophical vision. Moreover, this passage contains evidence that the paradigm of poetic dwelling was originally cast in terms of *Dasein*'s temporal existence in which our kind of existence is thrown equiprimordially into the three facets of time. The consequence for poetic dwelling is that the poet (construed as the paradigm for human existence) is cast into an in-between state; between humanity and the gods where the poet serves as the mediator that enables the institution of culture and the civil world. As Heidegger puts it, poetry as "the most innocent of all enterprises" requires that the poet be thrown out from the usual fare of the day *(aus dem Gewöhnlichen des Tages*). Rising above the herd mentality or the "they" was also an essential facet of *eigentlich existenz* in *Being and Time.* In his analysis of Hölderlin, Heidegger has assigned this obligation to the poet as cultural mediator. This also meant that *Dasein* was not simply concerned with the past as artifact, or the present as a mere now, but was, more importantly, comported toward the future in the movement Heidegger called *Wiederholung.* "Dwelling poetically" and *Dasein,* then, are both heralds of history.

Heidegger's analysis of Hölderlin in this passage claims the same mythical territory as does Vico's postulate of the originary poetic wisdom of humanity. In both cases poetry is credited with the founding of nations, and thus the emergence of the historical world. Moreover, poetry is the essence of the historical according to Heidegger, since the poet functions in the time in which the gods have abandoned the world and the future gods are yet to appear. The poet gives expression to this barren state and anticipates a new order or time. In doing so, the existence of the poet is thoroughly historical because it is in advance of history. In this isolated state, cast between the historical world and the gods, the circumspective vision of the poet longs for the past and looks to the future. Responding to the gods who speak through signs, poetic dwelling in the in-between intercepts divine signals and perceives in these first signs a completion that is yet to come.[28] Composed of both a circumspective vision, facing both the past and the future in an in-between state, and comported toward a totality on the basis of initial signs, Heidegger's analysis of poetic dwelling is a capstone to his historical analysis of *Dasein.*

Dwelling poetically and *Dasein* are the same historical structure. The mission of the poet, when thrown into the in-between, is to mediate the relationship between the signs of the gods and the "voice of the people" (Hölderlin's expression for the kind of tales in which a people is mindful of its relationship to the totality of beings). This thrown state, which Heidegger pointed to in his use of the term *ek-sistenz* in *Being and Time*, means that the poet, like *Dasein*, is situated between the past as having-been and the future as anticipated completion. The past is still present in the form of signs from the gods that are intercepted by the poets; and, on the basis of these signs, the poet, from the special vantage point of the in-between, anticipates a completion that enables the poet to foretell what is yet to be fulfilled. The poet, on these conditions, does not have to ask when a poem is complete. Completion comes with the initial reading of the signs from the gods; completion attends the original inspiration. In the same way, *Dasein* does not have to generate a completion, the manner in which *Dasein* is historically stretched along already constitutes a totality.

The poet, like *Dasein*, is a being thrown out of its place and into the in-between, between the gods and humanity. Like *Dasein,* from the vantage point of the *Augenblick* (the temporal in-between in *Being and Time*), the poet resolves what humanity is to become and how existence (*Dasein*) is to take up its abode.[29] Heidegger's refrain, here, is "poetically dwells humanity on this earth." The poet's mission is no less grave than the primordial responsibility of *Dasein* who is ontologically graced with resolutely attending to completion.

How this historical-poetic structure defines the limits of human existence is a question that neither Heidegger nor Hölderlin outlines in detail. What is spelled out, however, is the manner in which the prophetic-poetic voice of historical existence brings what is yet to be into the midst of our lives. This grasp of the future is not mediated by knowledge but through cultivating the "thinking heart" with a "humble patience" which releases us from "willing" and consequently releases us onto the "truth intact." This truth, which can only be realized after it resides in silence, ultimately manifests itself as a need. This structure is illuminated, once again, in the advice that Rilke gave to the young poet in a letter dated 12 August 1904. This advice, I believe, is a description of the way that the poetic-historical structure, of which Heidegger and Hölderlin spoke, transforms our lives. Rilke, then, helps us think Heidegger further: In counselling the poet on the perplexity we encounter from tribulations in our lives, Rilke explains the mysterious way in which destiny moves our lives. At difficult emotional times, Rilke advised, we stand perplexed and alienated because a trauma has distracted us from what we are accustomed to and intimate with (it removes us from the now). We stand in a moment of transition in which even the sorrow which has entered our hearts is concealed from us. This new thing, however, does not disappear, it is merely hidden; and, though it appears that nothing has happened, we have been changed in the way that a house is changed when a guest arrives.[30]

For Rilke, this is precisely the way that the future enters into us and transforms itself in us before it actually materializes. One shouldn't avoid these moments, nor should one indulge in the loneliness that accompanies them. One ought to be attentive to these barren moments, for this is the temporal juncture in which we become internally acquainted with our destiny. Rilke advised the young poet to become still and patient at these moments so that he might make what has entered into him more deeply his own; so that when in some "future" day it "happens" (publicly manifests itself) it will not appear strange and alienating. When the "future" finally "happens" it will strike us as necessity, since it has belonged to us for some time. Rilke maintained that the things that are really so akin to us, like death, the spirit world and even God, are forced out of life, and consequently the instincts that grasp these matters dissipate. Rilke counselled the young poet to re-evaluate the conventional conception of development in the way that science has always questioned customary concepts of motion, since destiny really emerges from within us, it does not whimsically enter us from the outside. On these terms, when the "future" finally "happens" it is really a former possibility that has realized itself. This new found sense of motion would exclude nothing from our lives and would prepare us for the most enigmatic of experiences. Rilke did not advise the young poet to embrace the occult or indulge in mysticism; he tried to comfort him with the realization that we take up our abode in a friendly atmosphere, in a domain that tallies with our existence. Moreover, after a long history of integration we find that we are so like what surrounds us that nothing stands against us.

In urging the young poet to trust even the most difficult of times, Rilke explained that there was nothing really in the world to mistrust; there were no truly alienating experiences, except the way of life that crowds out those experiences with which we are already most intimate. In comforting the young poet, Rilke was recommending the model of the artistic life which thrives even on melancholy and solitude. These things are not to be thwarted, he counselled, since with everything that comes ahead of you, they are secretly but resolutely at work in making everything uniquely your own. (Letter Ten, Paris the day after Christmas, 1908) Rilke warned the young poet that covering up the vicissitudes of life only prepares you for quasi-artistic professions. These professions, he had criticism, journalism and much of what stands for literature in mind, feign an allegiance with the arts while really subverting them. Falling prey to the parasitic world of the half-artistic professions, Rilke cautioned, eclipses the real experiences that unwittingly prepare us for life; they undercut that enduring patience (*Inständigkeit*), which is out of place in the hectic pace of ordinary life, but which really shapes us by putting us in touch with what surpasses us. [31]

# Notes

## Introduction

1. René Descartes, *Discourse on Method* (London: Penguin, 1968), p. 54.
2. G.W.F. Hegel, *Phenomenology of Mind,* translated by J.B. Baillie, (London: The MacMillan Co. 1949), p. 250. Karl Marx, "The Economic and Philosophic Manuscripts of 184," *Marx and Engels Reader*, Robert Tucker, editor, (New York: W.W. Norton, 2nd edition), p. 101.
4. Tucker, p. 105.
5. Tucker, p. 102.
6. Tucker, p. 105.
7. Richard Bernstein, *Beyond Objectivism and Relativism* (Philadelphia: University of Pennsylvania Press, 1985).
8. Aristotle, *Nichomachean Ethics*, translated by Martin Ostwald (The Bobbs-Merrill Company Inc. 1962, Indianapolis) Bk. 17, p. 14.
9. E. O. Wilson, *Sociobiology- The New Synthesis* (Cambridge, MA: Harvard University Press, 1975) reprinted in *Darwin, A Norton Critical Edition*, (New York, W.W. Norton and Co., 1979), p. 458. Wilson repeats himself in his latest consilience thesis. It is yet another form of biological reductionism that undermines other disciplines.
10. Wilson, p. 458.
11. Vico, Giambattista, *L'Autobiografia, il Cartegio, e le Poesie varie*, F. Nicoloni and B. Croce, editors (2nd edition; Bari: Laterza, 1929), pp. 13–14, quoted in Elio Gianturco's introduction to Vico's *On the Study Methods of our Time* (Cornell University Press, Ithaca and London, 1990), p. xxiii.
12. See Terrence Ball's article in *Vico and Marx,* Giorgio Tagliacozzo, editor (Atlantic Highlands, NJ: Humanities Press 1983), pp. 78–93, in which he cautions the liberal linking of Vico and Marx. He warns the critical reader that Vico and Marx meant two entirely different things when they referred to the priority of the creative transformation of reality. Ball's thesis admonishes the reader who casually makes the connection between Marx's emphasis of *praxis* as the primary tool in overcoming the unrelenting alienation of nature and the political/economic world and Vico's postulate of the primordial poetic (as productive) apprehension since Vico intended a kind of making that did not entail the actual production of a material object but the making of meaning. Ball's distinction does not annul the comparison between Marx and Vico on the grounds of the primacy of the "making" mode of human existence, it is a reproof for those that indulge in the anachronistic imposition of Vico onto Marx and vice versa. If Ball's thesis dissuades the reader from seeing the vital link between Vico and Marx, his fine tuning is only obfuscatory.
13. All references to Vico's *New Science* are from the Bergin and Fisch translation, (Ithaca, NY: Cornell University Press, 1986).
14. Donald Verene, *Vico's Science of the Imagination*, (Ithaca, NY: Cornell University Press, 1981), p. 80.

15. Verene, p. 81.
16. Verene, p. 83.
17. Verene, p. 88, ftn. 57.
18. Martin Heidegger, *Poetry, Language Thought*, translated by Alfred Hofstadter, (New York, Harper & Row, 1971) p. 215.
19. There are numerous studies linking Dante with Vico. Glauco Cambon's essay "Vico and Dante" in Tagliacozzo's *Giambattista Vico An International Symposium* (Baltimore, MD: Johns Hopkins University Press, 1969) reviews the history of this literature. Theophil Spoeri's study, *Dante und die europaische Literatur–Das Bild des Menschen in der Struktur des Sprache*, (Stuttgart, W. Korlemann, 1963), Cambon argues, demonstrates the most fundamental link between Dante and Vico. Spoeri enthusiastically equates Vico's primordial wood (within which the ontological space of the *luci* are cleared) with Dante's dark wood of error in which he encounters the three beasts that represent the three general categories of sin in *The Inferno*. Cambon adds that the two images are representative of what he calls the Fortunate Fall. [p.17] It is unclear what is meant by the expression "Fortunate Fall" but if Cambon means that Vico intended that the prototypical humans or giants of which he spoke began to ascend after they fell, then I would agree with the use of such a phrase. Dante's plight in the Dark Wood, however, is so fraught with theological doctrine that it is stretching it to link his sketch of the wilderness and the dark wood with Vico's picture of the primeval woods where the first "sublime poets" cleared the primordial locus that would serve as the ontological stage for the expression of poetic wisdom. I am arguing, however, that Dante's avowed dependence on the lasting impact upon his memory to assist him in the telling of his story indicates that he, like Vico, was steeped in rhetoric's cultivation of memory of place in the perfection of narrative discourse. Also, see Canto XIV, line 75 for a further example of how memory of a place can induce a physical response; in this case Dante experiences a shudder (literally a blood curdling response) from the memory of a river that ran deep red from the wood of the suicides (*lo cui rossore ancor mi raccapriccia*).
20. NS, 698. Also see Grassi's argument in "Critical or Topical Philosophy?", p. 48, *Vico An International Symposium,* Tagliacozzo, ed. which says that Aristotle's interpretation of topics does not illuminate the relationship of topics to philosophy as Vico's does. The strength of Vico's position, Grassi argues, lies in the understanding that the original premises of persuasive discourse are not deducible since they are the foundation for the logical process of presenting rational discourse. They are not deducible or subject to judgment because they are spontaneously invented within specific contexts that generate an array of *topoi* particular to the oratorical needs of the moment.
21. Grassi, Ernesto, "Vico, Marx and Heidegger, *Vico and Marx,* p. 246, edited by Giorgio Tagliacozzo, (Atlantic Highlands, NJ: Humanities Press, 1983).
22. Grassi, p. 240.
23. Heidegger, Martin, "Letter on Humanism," Basic Writings, edited by David Krell, (New York, Harper & Row, 1976), p. 235.
24. Krell, p. 235.
25. Krell, p. 222.
26. Krell, p. 218.
27. Krell, p. 213.
28. Krell, p. 237
29. Krell, p. 221.
30. Krell, p. 222.
31. Avineri, Shlomo, *The Social and Political Thought of Karl Marx,* (Cambridge: Cambridge University Press, 1967), p. 138.

32. Troeltsch's *Der Historismus und seine Probleme* was published in 1923, and it is comprised of essays and lectures that appeared in journals, particularly the *Historische Zeitschrift* from 1916 on. It is volume III in his collected works. I have used the Scientia Verlag Aalen 1977 edition. Hereafter it will be referred to as DH in the body of the text. Troeltsch reserved the designation *Kulturwissenschaften,* following Hermann and Rickert, for the interpretation of value theory. He used the general expression *Geisteswissenschaften* to designate the social sciences in their entirety as misleading. Troeltsch's logic of history embraced a sense of time that he characterized as a "living organic fusion of an instinctive continuity of meaning" which through "memory the past and present could be creatively united at every moment." (DH, p. 58) This fusion or *Verschmolzenheit,* as he called it, anticipated Gadamer's "fusion of horizons" for Troeltsch clearly characterized historical time as a stream in which the past and present are assumed in a productive manner. (DH, p.57) Gadamer's appropriation of Aristotle's concept of *phronesis,* that form of knowledge that is activated through choice and not mere know how (*techn*) or derived from universal principles (*epistem*) to distinguish the ethical impulse in philosophical hermeneutics corresponds to Troeltsch's logic of history in which he argued against mere theoretical and contemplative history in favor of historical understanding that always runs into the future. (p. 70, *Der Historismus)* Troeltsch dubbed his peculiar rendering of historical understanding as "historical seeing" (DH, p. 59) by which he meant a unifying life process (*ein einheitlicher Lebensprozess)* through which historical understanding is always affected by the "circle of culture surrounding the historical spectator and connecting him in actual consequence and effect." *Wirkungszusammenhängen* (DH, p. 75) Once again, we can see a clear connection with another Gadamerian principle, the concept of effective-history. Troeltsch, before Heidegger and Gadamer, had concluded that the structure of history entails a life process through which the past, present and future are mutually created from the "own standpoint" of the historical observer. Because of the vital position of one's own situation, the logic of history, for Troeltsch, entailed a material philosophy of history. A second volume on philosophy and history was to outline a material philosophy of history, but Troeltsch's early death denied this possibility. It is questionable, many Troeltsch scholars contend, if such a project was possible in light of the logic of history in *Der Historismus und seine Probleme.*
33. DH, p. 113.
34. DH, p. 78. This self-reflexive moment in historical interpretation was already suggested by nineteenth-century German philosophical historians like Gustav Droysen. The influence upon Heidegger and Gadamer from Droysen and other nineteenth century interpretations of history will become clear in the body of this work.

## Chapter 1

1. Boyd Shafer, "History, Not Art, Not Science, but History: Meanings and Uses of History." *The Vital Past*, edited by Steven Vaughn, (Athens, GA: University of Georgia Press), 1985. Shafer's thesis reflects the neo-Kantian challenge to the dominance of positivistic methodology.
2. Carl Becker, "Everyman His Own Historian," reprinted in *The Vital Past*, pp. 20–36.
3. In the recent past there have been numerous anthologies on Vico produced by Tagliacozzo and those dedicated to reviving Vico's philosophy. See especially *Vico: Past and Present*, (Atlantic Highlands, NJ: Humanities Press, 1981) and *Giambattista Vico, An International Symposium* (Baltimore, MD: The Johns Hopkins University

Press, 1969). The series *New Vico Studies* (Prometheus Books) edited by Giorgio Tagliacozzo and Donald Verene continues to disseminate Vichian scholarship. Vico continues to be at the center of the challenge to forms of philosophical expression that are devoid of historical interest. Generally speaking, it is primarily the analytical school of philosophy with its scientific pretensions that is most inimical to incorporating the historical. Vico proposed an alternative form of metaphysics with an emphasis on the poetic character of human discourse to counter the logocentric brand of metaphysics associated with Descartes. He argued in the *New Science* that human existence is primordially seated in a poetic transformation of reality that could equally serve as the foundation for a metaphysics since it too was engaged in apprehending the general from the vantage point of familiar particulars usually associated with the immediate sensuous awareness of the self.

4. Isaiah Berlin, *Vico and Herder, Two Studies in the History of Ideas*, (London: The Hogarth Press), 1976.
5. Vico, Giambattista, *New Science,* translated by Thomas Goddard Bergin and Max Harold Fisch, (Ithaca, NY: Cornell University Press, 1986), paragraph 405. In this section Vico establishes his poetic metaphysics as a more living alternative to rational metaphysics with its claim that man becomes all things by rationally comprehending them. Vico argues that the process of apprehending reality in rational metaphysics ought to be reversed to account for the nature of external reality which is totally different from the reality of human existence. Vico argued that it is more likely the case that man is incapable of really understanding all things and consequently apprehends reality on the basis of projecting himself onto reality and therefore transforming reality into himself. This sensuous transformation of reality, of course, is based on the more fundamental principle, expressed earlier in the *New Science,* that the civil world corresponds to modifications of the human mind and thus can only be understood through an examination of the mind. Both these principles share the belief that the apprehension of reality really arises from the sensuous mediation of human existence which overcomes the problems inherent in comprehending things that are totally external to the self. This amounts to Vico's solution to the learner's paradox that also afflicts Plato's epistemology, that is, how is it that learning can commence if you begin in a state of ignorance, not knowing that which you seek, otherwise why the search. Plato's solution is well-known and requires the positing of the immortality of the soul. Vico's solution to the problem establishes the preeminence of the mediation of the sensuous self; the suppressed premise in this thinking is, of course, that we are, all things being equal, in a state of constant acquaintance with our selves. All understanding, then, is predicated on transformation of reality into the familiar, the body. In this respect, Vico argues that metonymy and synechdoche, since they both involve obvious associations with the body or parts of the body with which, following his logic, the prototypical humans were most familiar, stand as the ground of abstract thinking though they are clearly ideas that have their origin in the sensuous. Though they have their origin in the sensuous apprehension of reality they are ideas just the same; serving the philosophic purpose of referring to the general or the whole from a familiarty with the particular, as is certainly the case with synecdoche.
6. Ibid., 348.
7. Ibid., 349.
8. Ibid., 143.
9. Ibid., 146.
10. Ibid., 138.
11. Ibid., paragraphs 129–131.
12. *Das Kapital*, book I, part IV, chapter 13 in a note numbered 89.

13. Grassi, Ernesto, "Critical or Topical Philosophy?" *Giambattista Vico, An International Symposium*, Giorgio Tagliacozzo, editor, (Baltimore, MD: The Johns Hopkins University Press), 1969.
14. Marx, Karl, "Contribution to the Critique of Hegel's *Philosophy of Right*," *The Marx-Engels Reader*, edited by Robert C. Tucker, second edition (New York: W. W. Norton and Co.), 1978, p. 64.
15. Hegel, G.W.F., *Philosophy of Right,* translated by S. W. Dyde, (London, George Bell and Sons) 1896, page xxvii of the preface.
16. Kierkegaard, Soren *Concluding Unscientific Postscript*, translated by David F. Swenson and Walter Lowrie, (Princeton, NJ: Princeton University Press), 1941, p. 110.
17. Ibid., p. 110.
18. Ibid., p. 111.
19. Piovanni, Pietro "Vico Without Hegel," *Vico: An International Symposium*, edited by Tagliacozzo, (Baltimore, MD: The Johns Hopkins University Press), 1969, p. 115.
20. Marx, "Contribution to the Critique of Hegel's *Philosophy of Right*, *The Marx-Engels Reader*, p. 53.
21. Marx, "The German Ideology,"*The Marx-Engels Reader*, p. 150.
22. Ibid., p. 163.
23. Marx, "The Communist Manifesto," The Marx-Engels Reader, p. 475.
24. Ibid., p. 487.
25. Heidegger's *Vom Wesen des Grundes* (1928) discusses the tendency in Western metaphysics to identify truth with propositional coherence. His intention in this essay was to introduce the disclosedness (*Erschlossenheit)* of *Dasein* as a more primordial condition of ontological truth. This condition creates a relationship with beings within which *Dasein* always perceives Being, thus indicating that the ontological difference, being able to distinguish between ontical and ontological truth, is implanted in the essence of *Dasein*, pp. 20-23 of the dual German-English edition, translated by Terrence Malick, (Chicago: Northwestern University Press 1969).
26. Ibid., p. 40.
27. Heidegger, Martin, "The Question Concerning Technology," in *The Question Concerning Technology and Other Essays,* translated by William Lovitt, (New York: Harper Torchbooks), 1977, pp. 13–14.
28. Ibid., p. 27.
29. Grassi, Ernesto, *Heidegger and the Question of Renaissance Humanism,* Center for Medieval and Renaissance Texts & Studies, (Binghamton, NY), 1983, p. 18. Now in Arizona State University.
30. Heidegger, Martin, *Being and Time,* translated by John Macquarrie and Edward Robinson, (New York: Harper and Row), p. 278; hereafter referred to as BT.
31. BT, p.287.
32. BT, p. 306.
33. BT, p. 308.
34. BT, p. 310.
35. BT, p. 355..
36. BT, p. 358.
37. BT, p. 372.
38. BT, p. 374..
39. BT, p. 375.
40. BT, p. 382..
41. BT, p. 242.

## Chapter 2

1. For a more comprehensive look at the German Historical School see Georg Iggers' thorough account of German history in The German Conception of History, The National Tradition of Historical Thought from Herder to the Present (Wesleyan University Press, Middletown CT: Wesleyan Press, 1968).
2. Hans Georg Gadamer, *Truth and Method*, (New York: PUB? 1982), p. 234.
3. See Karl Mannheim's argument in Ideology and Utopia, (New York, Meridan Press, 1936), translated by Louis Wirth and Edward Shils, pp.75–78, in which he clarifies historicism as a form of relationism. His intention was clearly to distinguish historicism from relativism, and in the process indicates the subjective foundations of historical knowledge.
4. Fritz Stern, editor, *The Varieties of History,* History as an Academic Discipline," (Cleveland: PUB? 1956), pp.170–177.
5. Friedrich Meinecke, "Kausalitäten und Werte in der Geschichte," Band IV, *Werke,* (Stuttgart: Koehler Verlag, 1965), p.60.
6. Ernst Troeltsch, *Der Historismus und seine Probleme,*(Scientia Verlag Aalen, 1977), p.75.
7. See Rudolff Makkreel's introduction to Dilthey's *Poetry and Experience,* (Princeton, NJ: Princeton University Press, 1985), pp.8–9, and Makkreel's book *Dilthey: Philosopher of the Human Studies,* "The Laws of the Imaginative Metamorphosis," pp. 90–109.
8. DH, p.169.
9. DH, p.66.
10. Meinecke, "Kausalitäten und Werte," p.64.
11. Meinecke, p.66.
12. Eduard Meyer, *Zur Theorie und Methodik der Geschichte,* 1902. This source is quoted in Troeltsch's *Der Historismus,* p.32.
13. Tillich's review first appeared in the *Theologische Literaturzeitung,* XIIX, 1924, pp. 25–30. Tillich's comments are praiseworthy overall, and he is keen to cite the innovations in Troeltsch's philosophy of history, especially the category of individual totality which challenges the Hegelian concept of universal history. Among the other reviewers of Troeltsch's book, Otto Hintze was the most critical. He argued that Troeltsch had conflated philosophical history with historiographical issues. This claim misses the point that these two aspects of history are compatible and even arise mutually. Troeltsch, like Heidegger, intended to establish the existential source of doing history. Troeltsch's contention that we cannot get out of our "historical skins" implies that to think is to think historically or about history. Meinecke and Mannheim also reviewed Troeltsch's book and both were clearly influenced by his thought. See especially Mannheim's *Ideology and Utopia*, pp. 75–79 and *Essays on the Sociology of Knowledge* (Oxford: Oxford University Press, 1952), pp. 97–124.
14. Meinecke, "Kausalitaeten und Werte," p.66.
15. Tillich's review of Troeltsch's *Der Historismus*, p.25.
16. Benedetto Croce, *History, Its Theory and Practice*, translated by Douglas Ainslie, (New York, Russell & Russell Publishers), p.62.
17. DH, p.36.
18. DH, p.37.
19. Troeltsch, "Empiricism and Platonism in the Philosophy of Religion," *Harvard Theological Review*, volume V, number 4, October 1912, pp. 402–422.
20. DH, p.48.

21. DH, p.48.
22. DH, p.48.
23. DH, p.51.
24. See Nietzsche's *The Use and Abuse of History, The Complete Works of Nietzsche*, edited by Oscar Levy and translated by Collins, (London: Allen & Unwin, 1909), Volume 5, p.1, for his allusion to the historicism in the person of Goethe. A similar opinion is expressed by Meinecke in *Die Entstehung des Historismus*, see especially the opening lines to the chapter on Goethe which expresses the sincere belief that without Goethe "we would not be where we are today."
25. See footnote 323 in *Der Historismus* for Troeltsch's reference to Alois Riehl's "Zur Einführung d. Gegenwart," 1919, p.222. It is Troeltsch's reiteration of the theoretical framework found in Sybel's preface to the *Historische Zeitschrift* (1859).
26. DH, p.58.
27. DH, p.59.
28. DH, p.76.
29. DH, p.77.
30. DH, p.77.
31. DH, p.78.
32. See G.P. Gooch's account of Ranke in *History and Historians in the Nineteenth Century* (first published in 1913) for details of Ranke's position in German historiography. Also see Georg Iggers's article "The Image of Ranke in America and Europe" in *History and Theory*, Volume XIV, Number 4, Beiheft 14, for a history of Ranke's ambiguous status in historical circles. For example, Iggers points out that when Ranke was imported into American historical circles his thinking was detached from the philosophical background of idealism, pp.17–40.
33. Benedetto Croce, who was also steeped in idealistic philosophy of history, argued fervently for the reciprocity of history and philosophy. The nineteenth-century innovation of historicizing philosophy meant that there was an indissoluble link between thought and history. Furthermore, it signified that genuine history depended on the fact that whatever event or deed is recalled must first "vibrate in the soul of the historian." (Croce, *History, Its Theory and Practice*, p.12) Equally moved by Vico's original design of the world of history in contrast to the world of nature, Croce insisted that philosophy of history was always resistant to external determinism. (*History, Its Theory and Practice*, p.64) The unity between (historical) deed and (historical) narrative is a living expression of the will of the historian and the enduring effects of the past. The collection of facts and the identity of pure causes in history is an insufficient method, since this approach can easily lead to a construal of the world as indefinitely linked in one causal chain. The historian must renounce the category of pure cause in favor of the opposite– an end in the transcendental sense. (*History, Its Theory and Practice*, p.67) The living fusion of the past and the present, making genuine history contemporaneous, directs the historian beyond past deed and present assimilation to future formation. Philosophy of history, on these terms, runs into ethics, and so Troeltsch's rendering of the *Geisteswissenschaften* as the historical-ethical sciences is fitting (DH, p.84).
34. Martin Heidegger, "Der Zeitbegriff in der Geschichtswissenschaften," *Zeitschrift für Philosophie und philosophische Kritik,* CLXI, 1916, pp.173–188.
35. Heidegger, p.174.
36. Heidegger, p.179.
37. Heidegger, p.179.
38. Martin Heidegger, *Sein und Zeit,* Band 2, Gesamtausgabe, (Frankfurt am Main: Klostermann, 1977), p.395.
39. BT, pp. 279–310.

40. BT, p. 369
41. BT, p. 380.
42. BT, p. 381.
43. BT, p. 431.
44. BT, p. 453 from the correspondence between Dilthey and Yorck.
45. David Hoy, "History, Historicity and Historiography in *Being and Time*", in *Heidegger and Modern Philosophy*, (New Haven, CT: Yale University Press, 1978) pp. 345–346.

## Chapter 3

1. Ernst Cassirer, *The Philosophy of Symbolic Forms, Mythical Thought*, translated by Ralph Manheim, (New Haven, CT: Yale University Press, 1955), p. 1.
2. Ibid., p. 105.
3. Mircea Eliade, *The Sacred and the Profane*, translated by Willard R. Trask, (New York: Harper & Row, 1957) p. 60. Eliade's use of the expression Parmenidean time suggests a parallel between Parmenides's rationalization that temporal delineation is illusory and mythic consciousness's intuition that time is internally identical and undifferentiated. Parmenides's demonstration followed the strictures of logical necessity; the pursuit of the *logos* showed that coming into being was a logical impossibility. Mythic consciousness, on the other hand, overcomes temporal determinations in the eternal return to origins, the affirmation that beginnings and coming into being underpins and constitutes human existence.
4. Cassirer, p. 111.
5. Mircea Eliade, *Myth and Reality*, (New York: Harper Colophan ??Books1975), p. 96. Myths are not restricted to stories of the sacred breaking into the profane, nor are they limited to narrative structure as is made clear in the anthropological approach to myth which derives from the analysis of myths within the context of living culture. Malinowski emphasized this facet of myth in his theory and directly informs Eliade's interpretation of myth as lived reality along the lines of the temporally dynamic conceptualization of history set forth in these pages. Though all myths do not refer to the activity of the Gods in a primordial time, they are all linked on the basis of their respective explanatory agendas, no matter how remote from the modern rigor of science. Eliade's theory is singled out here since his comprehensive approach combines the comparative approach to religious traditions, the structural anthropological interpretation of the archaic mind set, as well as the philological approach to myth with its emphasis on what Ernst Cassirer identified as the internal relation between name and object. Consequently, his interpretation of myth-making-consciousness is alert to the diachronic and cultural continuity of patterns of human behavior, especially as they are evident in the common tendency to recover the past or overcome time, as he put it, that universally characterizes mythical consciousness. Eliade's theory, above all, clarifies the historical consciousness that is explicit in mythical consciousness. His theory suggests that the archaic mindset is the prototypical temporal consciousness that is manifest as the historicist imagination in the nineteenth century. Eliade primarily emphasizes the connection between the "magical" power of myth in respect to the origin of cultural phenomenon and the healing power of recovering the past in psychoanalysis. Underlying both processes is a form of temporal transcendence that is made explicit in Heidegger's analysis of temporal consciousness in *Being and Time*. For his detailed argument of these connections see the chapter V of *Myth and Reality*, "Time Can Be Overcome." As the title suggests mythical consciousness and its later

day counterpart are forms of awareness that can transcend human, temporal finitude. Overall, Eliade's analysis of this temporal consciousness points to the salutary effects of achieving such a transcendence that is activated essentially through the powers of memory and recollection.

6. Eliade, p. 112.
7. Cassirer, p. 2.
8. Eliade, p. 113.
9. Eliade, p. 85. See also Eliade's *Myths, Dreams and Mysteries*, p. 49.
10. Eric Dardel, "The Mythic," *Sacred Narrative*, edited lan Dundes, (Berkeley: University of California Press, 1984), p. 232. The complete title of Dardel's essay is "The Mythic According to the Ethnological Work of Maurice Leenhardt." Dundes's notes to the essay trace Dardel's and Leenhardt's approach to myth to the philosophical tradition associated with Ernst Cassirer (*Language and Myth*, New York: Dover Publishers, Inc., 1946). Leenhardt's theory grew out of a long career in missionary work so it should be expected that his account would parallel the anthropological school associated with Malinowski.
11. BT, p. 237. This long hyphenated phrase represents the totality of the ontological structure of *Dasein*. This structural feature of *Dasein* signifies a solicitous attitude toward others and things-ready-to-hand in the environment. This latter attitude, obviously, entails the notion of time and implies the pre-eminence of the future in the constitution of *Dasein*. These two attitudes, however, are not unrelated, see note 29 below.
12. BT, p. 235.
13. David Couzens Hoy, "History, Historicity, and Historiography," *Heidegger* and *Modern Philosophy*, edited by Michael Murray, p. 336. Hoy argues that Heidegger's research into historical methodology is intimately connected to his subsequent 'destruction of ontology'. Hoy regards Heidegger's greatest achievement as the illumination of the historical nature of human existence. It follows that Heidegger denies the possibility that philosophy can begin with a presuppositionless starting point and culminate in ahistorical, eternal truths. Yorck, whom Heidegger depends on for his distinction between the ontical and the ontological, argued that philosophy cannot be foundationless, since it immediately refers to life. See pp. 449-455 in *Being and Time* for Heidegger's discussion of Yorck and Dilthey.
14. BT, p. 386.
15. Josiah Royce, *The World and the Individual*, (New York & London: The MacMillan Co., 1901), p. 125. The essays contained in this book were the Gifford Lectures delivered before the University of Aberdeen in 1900. In this particular lecture Royce defines our temporal form of experience as the form of the will and practical activity. He based his contention on the fact that the past and future are never idle in our minds but always stand in an effective relation to the present, thus constituting a whole. He expresses the same feeling of "ecstasy" that is conveyed in Heidegger's analysis of time consciousness.
16. Royce, p. 125.
17. Royce, p. 134.
18. Royce, p. 134.
19. Royce, p. 135.
20. Royce, p. 274. Royce's analysis of the Self, here, is from his lecture titled "The Human Self." What one ought to possess or create, according to Royce, is a principle that will ultimately distinguish a part of the world's life as our own. Royce argues for the possibility, in spite of our varied and transitory experiences, that a principle can emerge that will tie the seeming chaos of our experiences together.
21. Royce, p.275.

22. Royce, p.276. Defined as a task, human existence is not easily subsumed under traditional metaphysical categories. Like *Dasein*, whose essence is existence, Royce's view of the Self is existential in tone. Moreover, Royce's denial that the Self is the result of fulfilling a Soul-substance parallels Marx's definition of human essence in the VI thesis on Feuerbach; the human essence is not derivative of some transcendental substance that inheres in the individual. Feuerbach, Marx pointed out, abstracted the human essence from the historical process and consequently regarded the human essence in terms of an abstract "dumb generality" through which humans are connected. Feuerbach was not attentive to the historical process that gives rise to the conditions and social nexus that connect and define historical existence. Royce, Heidegger and especially Marx clearly resolve the human essence into a temporal concern conditioned by the application of will and projection through time.
23. John Dewey, *A Common Faith*, (New Haven, CT: Yale University Press, 1967), p. 41.
24. Dewey, p. 48
25. Dewey, p. 51.
26. BT, p. 310.
27. BT, p. 310.
28. BT, p. 310.
29. BT, p. 310. See note 11 that distinguishes Care as an attitude toward others and things-ready-to-hand and the characteristic of being-ahead-of-itself as typifying the attitude of *Dasein* toward itself. In this passage Heidegger demonstrates that these two attitudes are mutually arising.
30. BT, p.329.
31. BT, p. 332. This passage support's Hoy's understanding that Heidegger's research along the lines of historiography must be seen as a propaedeutic to the more profound investigation of ontology and historicity.
32. BT, p. 350.
33. BT, p. 365.
34. BT, p. 365.
35. BT, p. 365.
36. Karl Löwith, "Heidegger: Problem and Background of Existentialism," *Social Research*, vol. 15, 1948, p. 348.
37. Karl Löwith, "Nature, History, and Existentialism," *Social Research*, vol. 19, 1952, p.80.
38. Heidegger's expression *Geschehen* is translated by Macquarrie as "historizing," but since it applies to the special way *Dasein* happens historically, a more literal translation seems appropriate. Its fuller meaning will become clear when this special way of happening is analyzed. *Geschehen,* in its etymological kinship to *Geschichtlichke*it, which I have translated as historicity, might be rendered as historicizing to draw out its connection with historicized consciousness which was identified with *Dasein* earlier in this essay.
39. Heidegger's *Zeitbegriff* essay, *Zeitschrift für Philosophie und philosophische Kritik*, CLXI, p. 183.
40. Troeltsch refers to the so-called historicism predicated on the mathematical model, e.g., Comte's historical determinism, as "bad historicism," DH, p. 66.
41. Thomas Wren, "Heidegger's Philosophy of History," *Journal of the British Society for Phenomenology*, vol. III, no. 2, May 1972, p. 118.
42. Wren, p. 118.
43. Wren, p. 118.
44. BT, p. 346.
45. Ernesto Grassi, *Heidegger and the Question of Renaissance Humanism,* Medieval and Renaissance Texts and Studies, (New York: PUB?, 1983), p. 12.

46. BT, p. 374.
47. BT, p. 378.
48. BT, p. 378.
49. BT, p. 385.
50. BT, p. 385.
51. BT, p. 386.
52. BT, p. 386.
53. BT, p. 386.
54. Kierkegaard singles out Leibniz whose theory of monads teaches that "the whole of life is a repetition" based on the monad's perception of its subsequent entire development, the moment thus possesses the future in a calculated fashion. Troeltsch similarly relied on Leibniz' monadology, though with his adaptation of monadology to history the monads were not regarded as windowless.
55. Soren Kierkegaard, *Repetition*, (New York: Harper & Row, 1941), translated with an introduction by Walter Lowrie, p. 33.
56. BT, p. 397.
57. Kierkegaard, p. 33.
58. Kierkegaard, p. 34.
59. Kierkegaard, p. 52.
60. Kierkegaard, p. 39.
61. Kierkegaard, p. 39.
62. Kierkegaard, p. 90.
63. BT, p. 386. *Die Wiederholung macht dem Dasein seine eigene Geschichte erst offenbar*. Repetition obviously plays a crucial role in making *Dasein* aware of its historical nature.
64. BT, p. 390.
65. BT, p. 391.
66. BT, p. 391.
67. BT, p. 391.
68. We can recognize, here, the same dispute that Troeltsch had with Nietzsche in not formulating an organizing principle for life, that, for Troeltsch, is so crucial to rehabilitating historicism. Heidegger established the organizing principle with which one's temporal horizon is appropriated in regard to the "they" and everydayness.
69. BT, p. 397.
70. BT, p. 397. In this section Heidegger makes it clear that historiographical practice should follow from the historicity of *Dasein*. The science of history is determined by existential evidence, that is, the historical nature of *Dasein.*
71. BT, p. 397.
72. BT, p. 400. Troeltsch found more redeeming qualities in Ranke's theoretical approach to history. Heidegger, following the lead of Yorck, dismissed Ranke as hopelessly antiquarian.
73. BT, p. 401.
74. BT, p. 402. Yorck's feelings on this subject are quite clear. Philosophy in history is an immediate consequence to philosophizing and requires a break from traditional metaphysics.
75. BT, p. 402.
76. BT, p. 402. The quotes from Yorck that Heidegger relies on in paragraph 77 are taken from *Briefwechsel zwischen Wilhelm Dilthey und dem Grafen Yorck von Wartenburg*, 1877–1897, (Halle-an-der-Saale, 1923). At this particular point in the English translation by Macquarrie and Robinson, it should be noted, the translation of *Historie* changes from historiology to History with a capital "H."
77. BT, p. 310.

78. BT, p. 310.
79. Jean Paul Sartre, *Existentialism and Human Emotions* (New York: Philosophical Library,1957), pp. 60-67. This little book consists of selections from *Being and Nothingness* and *Existentialism.* This particular section, demonstrating the correspondence between God and man, is taken from *Being and Nothingness*, translated Hazel E. Barnes.

## Chapter 4

1. Donald Phillip Verene, *Hegel's Recollection: A Study of Images in the Phenomenology of Spirit*, (Albany, NY: SUNY Press), 1985.
2. G.W.F. Hegel, *Phenomenology of Spirit*, translated by J. B. Baillie, Second Edition, (New York: The Macmillan Company), 1931, p. 89 to the Preface
3. Ibid., p. 93 of the Preface.
4. Verene, p. 11.
5. Hans-Georg Gadamer, *Truth and Method*, (New York: Pub?, 1975), p.93. Hereafter cited as TM.
6. TM, p. 93.
7. TM, p. 97.
8. TM, p. 98.
9. TM, p. 109.
10. TM., p. 110.
11. TM., p. 250.
12. TM., p. 263.
13. TM., p. 263.
14. TM., p. 268.
15. TM., p. 268.
16. TM., p. 276.
17. TM., p. 273.
18. BT, p. 431.
19. TM, p. 320.
20. TM., p. 320.
21. TM, p. 323.
22. TM., p. 342.
23. Hans-Georg Gadamer, *Philosophical Apprenticeships*, "On the Origin of Philosophical Hermeneutics," translated by Robert R. Sullivan, (Cambridge, MA: The MIT Press, 1985), p. 183.
24. Aristotle's *Nichomachean Ethics*, Bk.VI, Chap.7, Lines 10–20.
25. Johann Gustav Droysen, *Outline of the Principles of History*, translated by E. Benjamin Andrews, (Boston: Beacon Press, 1893), Appendices II, "Nature and History," p. 91.
26. Ibid., p. 99.
27. G.W.F. Hegel, *Reason in History*, translated by Robert S. Hartman, (New York: Liberal Arts Press, 1953), p.5.
28. Ibid., p. 5.
29. Hegel's failure to appreciate the hermeneutic space in experience was addressed by the barrage of criticisms that followed in the nineteenth century. The rise of neo-Kantianism, in particular, was in many respects a refutation of Hegel's system. This critique was largely guided by the repeated assessment of Hegel's notorious dictum: "What is rational is real and what is real is rational." The call back to Kant in relation to this synoptic grasp of Hegel's system is comprehensively recorded in Herbert Schnaedelbach's history of this period, *Philosophy in Germany, 1831–1933*, translated by Eric Matthews, (Cambridge: Cambridge University Press, 1984).

30. Martin Heidegger, "Der Zeitbegriff in der Geschichtswissenschaften," *Frühe Schriften*, Band I, (Frankfurt: Vittorio Klostermann, 1978), p. 427.
31. Ibid., p. 427.
32. Martin Heidegger, "On the Origin of the Work of Art," *Poetry, Language, Thought*, translated by Albert Hofstadter (New York: Harper and Row, 1975), p. 48.
33. Martin Heidegger, *Unterwegs zur Sprache* (Pfullingen:Neske,1959), p. 193, quoted from Walter Biemel's "Poetry and Language in Heidegger," p. 88, *On Heidegger and Language*, edited by Joseph Kockelmans, (Chicago: Northwestern University Press, 1972).

## Chapter 5

1. *An Essay Concerning Human Understanding*, ed. A.C. Fraser,2 vols. (Oxford: Clarendon Press, 1894), 2 2:146, quoted in Donald Verene's "The New Art of Narration" in *New Vico Studies,* (Atlantic Highlands, NJ: Humanities Press, 1983), p. 22.
2. Martin Heidegger, *Brief über den "Humanismus"* in *Platons Lehre von der Wahrheit*, (Bern: Verlag A. Francke AG., 1947), p.53.
3. Rickert discusses his opposition to both Dilthey and Troeltsch in chapter 4 of his major work *The Limits of Concept Formation in the Natural Sciences*, edited and translated by Guy Oakes (Cambridge: Cambridge University Press, 1986). See section 9 of chapter 4, in which he distinguishes his approach to history from Dilthey's and Troeltsch's. He takes Dilthey to task for being too tentative and not systematic enough. Both Troeltsch and Dilthey failed to appreciate the distinction between the reality of historical creations, as in the case of art produced by artists, and the associated conceptualization of historical reality that Rickert identified as nonreal configurations of meaning. See especially pages 140–145. The concern of the historian ought to be with the real processes that give rise to these so-called configurations of meaning; these processes can only be conceptualized on the basis of the valuation and volition that generated the real historical process.
4. Donald Davidson, "The Method of Truth in Metaphysics," *Philosophy: End or Transformation*, (Cambridge, MA. The MIT Press, 1987), p. 168.
5. Ibid.
6. Heinrich Ott, "Hermeneutic and Personal Structure", *On Heidegger and Language*, edited by Joseph Kocklemans, (Chicago: Northwestern University Press, 1972). See the sections titled "The Hermeneutic Structure of Language" and "The Personal Structure of Language." Ott's argument against the insufficiency of the scientific approach to reveal the essence of language is cast in theological terms, but it, too, like the historical analogue mustered from Heinrich Rickert's work, does not dismiss the relevance of scientific conceptualization where it might apply. However, the scientific manner of analyzing language, since it regards language more as an ability which is simply an extension of human existence, does not address the true essence of language which is equiprimordial with human existence. Heidegger's position, more accurately, was that human existence dwells inside language making it more than a simple instrument in the pursuit of ends. The scientistic conception of language regards language as a way of imparting information and does not understand that language as communication calls things into their being or brings things to the fore, as Heidegger might put it.
7. Robert Caponigri, *Time and Idea*, (Notre Dame: Notre Dame University Press, 1953), p. 181.
8. Giambattista Vico "On the Ancient Wisdom of the Italians", III, *Vico: Selected Writings*, ed. and tr. by Leon Pompa, (Cambridge: Cambridge University Press, 1982), pp.58-59

9. Heidegger, *Brief über den "Humanismus,"* p. 54.
10. Horace, *The Art of Poetry*, line 59, translated by H. Rushton Fairclough (London: PUB? 1936), p. 457. Horace's claim parallels Vico's principle of the certainty provided by the *sensus communis.* Vico conceived the philological foundations of the *sensus communis* as the source of natural law, which in turn was determined by need and utility. (Principle 141 of the *New Science*) Both Horace and Vico, then, recognized the spontaneous creative powers of language to respond to meaning arising from practical contexts geared toward need and utility. Need and utility are not cognitively apprehended, nor are they sanctioned by rationally discursive methods; they are rooted in practice and application and consequently resist the impersonal rigors of grammar and logic.
11. Ernesto Grassi, *Rhetoric as Philosophy*, (Unioversity Park, PA: Penn State University Press, 1980 ), p. 76.
12. See George Vick's article in *Heidegger and Modern Philosophy*, ed., Michael Murray, (New Haven, CT: Yale University Press, 1978), pp. 210–213.
13. Wilhelm Dilthey, *Poetry and Experience*, Volume V of Selected Works, translated and edited by Rudolf A. Makkreel and Frithjof Rodi, (Princeton, NJ: Princeton University Press, 1985), p. 56
14. Alfred North Whitehead, *Adventures in Ideas*, (New York: The Macmillan Company, 1933), p. 226.
15. Ibid.
16. Heidegger, *Brief über den "Humanismus* ," p. 55.
17. Ibid., p. 54.
18. Ibid., p. 90.
19. Ibid., p. 91.
20. Heidegger, *Zur Erörterung der Gelassenheit*, (Pfullingen: Verlag Guenther Neske, 1959), p.46
21. Ibid., p. 48.
22. Ibid., p. 62.
23. Ibid., p. 64.
24. Ibid., p. 42.
25. Ibid., p. 42.
26. Ibid., p. 47.
27. Rainer Maria Rilke. *Letters to a Young Poet*, translated by M. D. Herter Norton, (New York: W. W. Norton and Co. 1962), p. 30.
28. Heidegger, *Erläuterungen zu Hölderlins Dichtung*, (Frankfurt am Main: Vittoria Klostermann, 1971), p. 46.
29. Ibid., p. 47.
30. Rainer Maria Rilke, *Briefe an einen jungen Dichter* (Leipzig: Insel-Verlag, 1929), pp. 44-45.
31. Rilke expresses these sentiments in a letter to Rodin from the same period as Letter Ten to the young poet. In it he acknowledges that Rodin has taught him the creative powers of the enduring patience of which Heidegger spoke in his dialogue on *Gelassenheit.* Heidegger, too, in the application of the expression *Inständigkeit,* recognized the role this lasting patience has in putting us in touch with that which transcends us and consequently defines us. It is no ordinary patience, but a tenacious resolve that enlists a non-willing disposition and a solitary stillness that accepts the constant unwitting movement toward destiny and that which surpasses us. The English edition of Rilke's letters, *Letters to a Young Poet*, translated by M.D. Herter Norton, (W.W. Norton & Co. 1934) contains a section titled "Chronicle, 1903–1908" in which this information is found, p. 123.

# Bibliography


Anderson, Chester, G., ed., *A Portrait of the Artist as a Young Man, Text, Criticism and Notes,* New York: Penguin Books, 1977.

Aristotle, *Nichomachean Ethics*, translated by Martin Ostwald, Indianapolis, IN: The Bobbs-Merrill Company Inc., 1962

Avineri, Shlomo, *The Social and Political Thought of Karl Marx,* Cambridge: Cambridge University Press, 1967.

Ball, Terrence, *Vico and Marx,* edited by Giorgio Tagliacozzo, Atlantic Highlands, NJ: Humanities Press, 1983.

Baynes, Kenneth, Bohman, James and McCarthy, Thomas, editors, *Philosophy, End or Transformation?* Cambridge, MA: The MIT Press, 1987.

Becker, Carl, "Everyman His Own Historian," American Historical Association Address, delivered at Minneapolis, December 29, 1931. *American Historical Review*, Volume 37, Issue 2, pp. 221-236.

Bergson, Henri, *Creative Evolution*, translated by Arthur Mitchell, New York: Henry Holt and Company, 1911.

Bergson, Henri, *The Creative Mind, An Introduction to Metaphysics,* Totiwa, NJ: Littlefield, Adams & Co.,1975.

Berlin, Isaiah, *Vico and Herder, Two Studies in the History of Ideas,* London: The Hogarth Press, 1976.

Bergin, Thomas G. & Fisch, Max H., Vico, Giambattista, *New Science,* Ithaca, NY: Cornell University Press, 1986.

Bernstein, Richard, *Beyond Objectivism and Relativism*, Philadelphia: University of Pennsylvania Press, 1985.

Biemel, Walter, "Poetry and Language in Heidegger, " *On Heidegger and Language*, edited by Joseph Kockelmans, Chicago: Northwestern University Press, 1972.

Buttigieg, Joseph, A., *A Portrait of the Artist in Different Perspective*, Athens, OH: Ohio University Press, 1987.

Caponigri, Robert, *Time and Idea*, Notre Dame: University of Notre Dame Press, 1953.

Caputo, John, D., *Demythologizing Heidegger,* Indianapolis: Indiana University Press, 1993.

Cassirer, Ernst, *The Philosophy of Symbolic Forms, Mythical Thought,* translated by Ralph Manheim, New Haven, CT: Yale University Press, 1955.

Cotton, James Harry, *Royce on the Human Self,* Cambridge, MA: Harvard University Press, 1954.

Croce, Benedetto, *History, Its Theory and Practice*, translated by Douglas Ainslie, New York: W.W. Norton, 1960.
Croce, Benedetto, *History and the Story of Liberty*, translated by Sylvia Sprigge, New York: W. W. Norton, 1941.
Cunningham, Gustav Watts, *A Study in the Philosophy of Bergson,* Longmans, New York: Green and Company, 1916.
Davidson, Donald, "The Method of Truth in Metaphysics," *Philosophy: End or Transformation*, Cambridge, MA: The MIT Press, 1987.
Descartes, René, *Discourse on Method*, London: Penguin Books,1968.
Dewey, John, *A Common Faith*, New Haven, CT: Yale University Press, 1967.
Dilthey, Wilhelm, *Poetry and Experience*, Volume V of Selected Works, translated and edited by Rudolf A. Makkreel and Frithjof Rodi, Princeton, NJ: Princeton University Press, 1985.
Dilthey, Wilhem, *Die Geistige Welt, Einleitung in die Philosophie des Lebens, Gesammelte Schriften VI. Band,* Vandenhoeck & Ruprecht, Göttingen , 1978
Droysen, Johann Gustav, *Grundriss der Historik.,* Halle a.Saale: Max Niemeyer, 1925; *Outline of the Principles of History*, translated by E. Benjamin Andrews, Boston, MA: Ginn Publishers, 1893.
Dundes, Alan, ed., *Sacred Narratives, Readings in the Theory of Myth*, Berkeley: University of California Press, 1984.
Eliade, Mircea, *The Sacred and the Profane*, translated by Willard R. Trask, New York: Harper & Row, 1957.
Eliade, Mircea, *Myth and Reality*, New York: Harper Colophon Books, 1975.
Farias, Victor, *Heidegger and Nazism,*edited by Joseph Margolis and Tom Rockmore, translated by Gabriel R. Ricci and Paul Burrell, Philadelphia: Temple University Press, 1989.
Fraisse, Paul, *The Psychology of Time*, translated by Jennifer Leith, New York: Harper & Row, 1963.
Gadamer, Hans Georg, *Truth and Method*, New York: The Crossroad Publishing Co.,1982.
Gadamer, Hans Georg, *Philosophical Apprenticeships*, translated by Robert R. Sullivan, Cambridge, MA: The MIT Press, 1985.
Gadamer, Hans-Georg, *Heidegger's Ways*, translated by John W. Stanley, Albany, NY: SUNY Press, 1994.
Gillespie, Michael, Allen, *Hegel, Heidegger, and the Ground of History*, Chicago: The University of Chicago Press, 1984.
Goethe, Johann, Wolfgang, von, *Faust*, Part I, translated with an introduction by Peter Salm, Toronto, New York: Bantam Books, 1985.
Goethe, Johann, Wolfgang, von, *Faust,* Part II, translated with an introduction by Philip Wayne, Middlesex, England: Penguin Books, 1986.
Gooch, G.P., *History and Historians in the Nineteenth Century* (first published in 1913), Boston: Beacon Press, 1959.
Grassi, Ernesto, "Vico, Marx and Heidegger, *Vico and Marx,* Giorgio Tagliacozzo editor, Atlantic Highlands, NJ: Humanities Press, 1983.
Grassi, Ernesto, "Critical or Topical Philosophy?" *Giambattista Vico, An International Symposium,* Giorgio Tagliacozzo, editor, Baltimore, MD: The Johns Hopkins Press, 1969.

Grassi, Ernesto, *Heidegger and the Question of Renaissance Humanism,* Binghamton, NY: Center for Medieval and Renaissance Texts & Studies, 1983.

Grassi, Ernesto, *Rhetoric as Philosophy*, University Park, PA: Penn State University Press, 1980.

Hegel, G.W.F., *Reason in History*, translated by Robert S. Hartman, New York: Liberal Arts Press, 1953.

Hegel, G.W.F., *Philosophy of Right,* translated by S. W. Dyde, London: George Bell and Sons, 1896.

Hegel, G.W.F. *Phenomenology of Spirit*, translated by J. B. Baillie, 2nd edition, New York: The Macmillan Company, 1931.

Heidegger, Martin, *Poetry, Language Thought*, translated by Alfred Hofstadter, New York: Harper & Row, 1971.

Heidegger, Martin, *Vom Wesen des Grundes*, translated by Terrence Malick, Chicago: Northwestern University Press 1969.

Heidegger, Martin, "The Question Concerning Technology," in *The Question Concerning Technology and Other Essays,* translated by William Lovitt, New York: Harper Torchbooks, 1977.

Heidegger, Martin, *Being and Time,* translated by John Macquarrie and Edward Robinson, New York: Harper & Row, 1962.

Heidegger, Martin, "Der Zeitbegriff in der Geschichtswissenschaften," *Zeitschrift für Philosophie und philosophische Kritik,* CLXI, 1916.

Heidegger, Martin, *Sein und Zeit,* Band 2, Gesamtausgabe, Frankfurt am Main: VIttorio Klostermann, 1977.

Heidegger, Martin, "Der Zeitbegriff in der Geschichtswissenschaften," *Frühe Schriften*, Band I, Frankfurt am Main: Vittorio Klostermann, 1978.

Heidegger, Martin, "On the Origin of the Work of Art," *Poetry, Language, Thought*, translated by Albert Hofstadter, New York: Harper & Row, 1975.

Heidegger, Martin, *Unterwegs zur Sprache*, Verlag Guenther Neske Pfulligen, 1959.

Heidegger, Martin, *Brief über den "Humanismus*" in *Platons Lehre von der Wahrheit*, Verlag A. Francke AG. Bern, 1947.

Heidegger, Martin, *Zur Erörterung der Gelassenheit*, Verlag Guenther Neske Pfullingen, 1959.

Heidegger, Martin, *Erläuterungen zu Hölderlins Dichtung*, Franffurt am Main: Vittoria Klostermann, 1971.

Heidegger, Martin, *Hegel's Concept of Experience,* translated by Kenley Royce Dove, New York: Harper & Row, 1970.

Heidegger, Martin, *Basic Writings*, edited with an introduction by David Farrel Krell, New York: Harper & Row, 1977.

Heller, Erich, *The Artist's Journey Into the Interior*, New York: Harcourt Brace Jovanovich, 1965.

Heller, Erich, *The Importance of Nietzsche*, Chicago: University of Chicago Press, 1988.

Horace, *Satires, Epistles, and Ars Poetica*, with an English translation by H. Rushton Fairclough, London: Wm. Heinemann; New York: G. P. Putnam's Sons, 1926.

Horkheimer, Max, *Critical Theory, Selected Essays,* translated by Matthew J. O'Connell, New York: Continuum, 1982.

Hoy, David Couzens "History, Historicity and Historiography in *Being and Time*", in *Heidegger and Modern Philosophy*, ed., Michael Murray, New Haven, CT: Yale University Press, 1978.

Hughes, H. Stuart, *Consciousness and Society*, New York: Vintage Books, 1977.

Husserl, Edmund, *The Phenomenology of Internal Time-Consciousness*, translated by James S. Churchill, introduction by Calvin O. Schrag, Indianapolis: Indiana University Press, 1964.

Iggers, Georg *The German Conception of History*, Middletown, CT: Wesleyan University Press, 1968.

Iggers, Georg "The Image of Ranke in America and Europe" in *History and Theory*, Volume XIV, Number 4, Beiheft 14.

Jay, Martin, *The Dialectical Imagination,* Berkeley: University of California Press, 1996,

Kierkegaard, Soren *Concluding Unscientific Postscript*, translated by David F. Swenson and Walter Lowrie, Princeton, NJ: Princeton University Press, 1941.

Kierkegaard, Soren, *Repetition*, translated with an introduction by Walter Lowrie, New York: Harper & Row, 1941.

Lewis, Wyndham, *Time and Western Man*, edited with afterword and notes by Paul Edwards, Santa Rosa: Black Sparrow Press, 1993.

Lewis, Wyndham, *Blasting and Bombardiering, An Autobiography (1914-1926)*, London: John Calder Publishing Co., 1982.

Lilla, Mark, *G.B. Vico, The Making of and Anti-Modern*, Cambridge, MA: Harvard University Press, 1993.

Löwith, Karl, "Heidegger: Problem and Background of Existentialism," *Social Research*, Vol. 15, 1948.

Löwith, Karl, "Nature, History, and Existentialism," *Social Research*, Vol. 19, 1952.

Makkreel, Rudolf, A., *Dilthey: Philosopher of the Human Studies,* Princeton, NJ: Princeton University Press, 1975.

Mannheim, Karl, *Ideology and Utopia*, translated by Louis Wirth and Edward Shils, New York: Harcourt Brace Jovanovich, 1936.

McKeon, Richard, *Rhetoric, Essays in Invention & Discovery*, edited with an Introduction by Mark Backman, Woodbridge, CT: Ox Bow Press,1987.

Meinecke, Friedrich, "Kausalitäten und Werte in der Geschichte," Band IV, *Werke,* Stuttgart: Koehler Verlag, 1965.

Meyer, Eduard, Zur *Theorie und Methodik der Geschichte,* Halle a.Saale: Max Niemeyer, 1902.

Murdoch, Iris, *The Fire & the Sun, Why Plato Banished the Artists*, Oxford: Clarendon Press, 1977.

Murray, Michael, editor, *Heidegger & Modern Philosophy*, New Haven, CT: Yale University Press, 1978.

Nietzsche, *The Use and Abuse of History, The Complete Works of Nietzsche*, Volume 5, edited by Oscar Levy and translated by Adrian Collins, London: Allen & Unwin, 1909.

Ott, Heinrich, "Hermeneutic and Personal Structure", *On Heidegger and Language*, edited by Joseph Kocklemans, Chicago: Northwestern University Press, 1972.

Piovanni, Pietro "Vico Without Hegel," *Vico: An International Symposium*, edited by Tagliacozzo, Baltimore, MD: The Johns Hopkins University Press, 1969.

Pompa, Leon, *Vico, A Study of the 'New Science',* Cambridge: Cambridge University Press, 1990.

Press, Gerald, A., *The Development of the Idea of History in Antiquity,* Montreal: McGill University Press, 1982.

Reist, Benjamin, A., *Toward a Theology of Involvement, The Thought of Ernst Troeltsch*, Phiadelphia: The Westminster Press, 1966.

Rickert, Heinrich, *The Limits of Concept Formation in the Natural Sciences*, edited and translated by Guy Oakes, Cambridge University Press, 1986.

Ricoeur, Paul, "The Task of Hermeneutics," *Heidegger and Modern Philosophy*, ed., Michael Murray, New Haven, CT: Yale University Press, 1978.

Rilke, Rainer Maria, *Briefe an einen jungen Dichter*, Insel-Verlag, Leipzig, 1929.

Rilke, Rainer Maria, *Letters to a Young Poet*, translated by M.D. Herter Norton, New York: W.W. Norton & Co., 1934.

Royce, Josiah, *The Sources of Religious Insight*, New York: Charles Scribner's Sons, 1912.

Royce, Josiah, *The Religious Aspects of Philosophy,* New York: Houghton Mifflin Company, 1885.

Royce, Josiah, *The World and the Individual*, New York: The Macmillan Co., 1901.

Schnaedelbach, Herbert, *Philosophy in Germany, 1831–1933*, translated by Eric Matthews, Cambridge: Cambridge University Press, 1984.

Simmel, Georg, *The Sociology of Georg Simmel,* translated with an introduction by Kurt H. Wolf, Glencoe, IL: The Free Press, 1950.

Shafer, Boyd, "History, Not Art, Not Science, but History: Meanings and Uses of History." *The Vital Past*, Steven Vaughn, ed., Athens: University of Georgia Press, 1985.

Spoeri, Theophil, *Dante und die europaische Literatur–Das Bild des Menschen in der Struktur des Sprache*, Stuttgart,: W. Kohlemann Verlag, 1963.

Stern, Fritz ed., *The Varieties of History,* New York: Meridian Books, 1956.

Sybel, Heinrich von, "Introduction", first issue of *Historische Zeitschrift* 1859.

Tagliacozzo, Giorgio, ed., *Vico: Past and Present,* Atlantic Highlands, NJ: Humanities Press, 1981.

Troeltsch, Ernst, *Der Historismus und seine Probleme,* Scientia Verlag Aalen, 1977.

Tillich, Paul, "Ernst Troeltsch: Historismus und seine Probleme", *Theologische Literaturzeitung,* XLIX, 1924. pp. 25-30.

Troeltsch, Tillich, "Empiricism and Platonism in the Philosophy of Religion," *Harvard Theological Review*, Volume V, Number 4, October 1912.

Tucker, Robert, editor, *Marx and Engels Reade* , 2nd edition, New York: W.W. Norton, 1978.

Verene. Donald, "The New Art of Narration" in *New Vico Studies,* Atlantic Highlands, NJ: Humanities Press, 1983.

Verene, Donald Phillip, *Hegel's Recollection: A Study of Images in the Phenomenology of Spirit*, Albany, NY: SUNY Press, 1985.

Verene, Donald, *Vico's Science of the Imagination*, Ithaca, NY: Cornell University Press, 1981.

Vick, George, "Heidegger's Linguistic Rehabilitation of Parmenides' "Being" ", *Heidegger and Modern Philosophy*, ed., Michael Murray, New Haven, CT: Yale University Press, 1978.

Vico, Giambattista, *L'Autobiografia, il Cartegio, e le Poesie varie*, eds. F. Nicoloni and B. Croce, 2nd edition; Bari: Laterza, 1929.

Vico, *New Science,* Thomas Bergin and Max Fisch translation, Ithaca, NY: Cornell University Press, 1986.

Vico, Giambattista, "On the Ancient Wisdom of the Italians", III, *Vico: Selected Writings*, ed. and tr. by Leon Pompa, Cambridge University Press, 1982.

Vico, Giambattista, *The Autobiography of Giambattista Vico*, translated from the Italian by Max Harold Fisch & Thomas Goddard Bergin, Ithaca, NY: Cornell University Press, 1993.

Vico, Giambattista, *On Humanistic Education (Six Inaugural Orations, 1699-1707),* with an introduction by Donald Phillip Verene, Ithaca, NY: Cornell University Press, 1993.

Vico, Giambattista, *On The Study Methods of Our Time*, translated with an introductioin and notes by Elio Gianturco, Ithaca, NY: Cornell University Press, 1990.

Wardy, Robert, *The Birth of Rhetoric, Gorgias, Plato and Their Successors*, London and New York: Routledge, 1996.

White, Hayden V., and Tagliocozzo, Giorgio, editors, *Vico, Giambattista, An International Symposium*, Baltimore, MD: The Johns Hopkins Press, 1969.

Whitehead, Alfred North, *Adventures in Ideas*, New York: The Macmillan Company, 1933.

Whitney, George Tapley, ed., *The Heritage of Kant*, Princeton, NJ: Princeton University Press, 1939.

Wilson, E. O., *Sociobiology- The New Synthesis* reprinted in *Darwin, A Norton Critical Edition*, New York, 1979.

Wolff, Robert, Pail, editor, *Kant, A Collection of Critical Essays*, Notre Dame, IN: University of Notre Dame Press, 1967.

Wren, Thomas "Heidegger's Philosophy of History," *The Journal of the British Society for Phenomenology*, Vol. III, No. 2, May 1972.

# Index